Economic Growth and Sustainability

Dedicated to the
next generations:
those who will find
the answers

Clark **Sloan **Amber **Beth **Vicki **Grant **Thomas **Debbie **Pierce **Susie **

Economic Growth and Sustainability

Systems Thinking for a Complex World

Karen L. Higgins

AMSTERDAM • BOSTON • HEIDELBERG • LONDON • NEW YORK
OXFORD • PARIS • SAN DIEGO • SAN FRANCISCO • SINGAPORE
SYDNEY • TOKYO

Academic Press is an Imprint of Elsevier

Academic Press is an imprint of Elsevier
525 B Street, Suite 1800, San Diego, CA 92101-4495, USA
32 Jamestown Road, London NW1 7BY, UK
225 Wyman Street, Waltham, MA 02451, USA

British Library Cataloguing-in-Publication Data
A catalogue record for this book is available from the British Library

Library of Congress Cataloging-in-Publication Data
A catalog record for this book is available from the Library of Congress

ISBN: 978-0-12-802204-7

For information on all Academic Press publications
visit our website at store.elsevier.com

Printed and bound in the United States of America

15 16 17 18 19   10 9 8 7 6 5 4 3 2 1

Working together
to grow libraries in
developing countries

www.elsevier.com • www.bookaid.org

Contents

Preface   xi
About the Author   xv
Introduction   xvii

1   The Secret's in the Overlap: Sustainability
    as an Integrated System   1

    1.1 Environmentalism and Sustainability   2
         1.1.1 The Sustainability Revolution   2
         1.1.2 Sustainability and Interdependence   4
    1.2 Systems Thinking   4
         1.2.1 Systems Thinking Constructs   5
         1.2.2 Boundaries and Limits   7
         1.2.3 Systems Thinking Applied   9
    1.3 Lessons for the Future   10
         References   11

2   Living in a Bubble: A Mental Model of How
    the World Works   13

    2.1 Defining Our Predominant Mental Model   14
         2.1.1 Constraints of Individual Mental Models   15
         2.1.2 Primary Beliefs in Our Mental Model   16
    2.2 Economic Growth and Human Thriving   16
         2.2.1 Individuals and Materialism   16
         2.2.2 Collectives and Economic Success   17
         2.2.3 World Economic Trends   18
    2.3 Energy and Technology Advances   18
         2.3.1 Availability of Energy   19
         2.3.2 Energy-Related Technology   19
    2.4 Population Growth and Pollution   21
         2.4.1 Population and Our Mental Model   21
         2.4.2 Pollution and Our Mental Model   22
         References   23

3   The Ant Who Lives Forever: A Systems
    Interpretation of Our Mental Model   25

    3.1 Systems Depiction of Our Mental Model   25
         3.1.1 Economic Growth and Personal Gratification   26
         3.1.2 Abundant Energy and Technology Advances   27
         3.1.3 Population and Pollution Are External   28

3.2   Implications of Our Mental Model   28
      3.2.1   Unbounded Growth   29
      3.2.2   Unrealistic Beliefs   29
      References   30

4   Addicted to Growth: Economic Growth
    Promises Happiness and Well-Being   31

    4.1   Addicted to Economic Growth   32
    4.2   Economic Growth Trends   33
          4.2.1   Population, GDP, and Consumer Spending Compared   33
          4.2.2   Implications for Standards of Living   35
          4.2.3   Short-Term Benefits   36
    4.3   Repercussions of Economic Growth   37
    4.4   A Promise Broken: Creating a New Perspective   38
    References   38

5   Two Faces of Happiness: Instant Gratification
    versus Sustainable Well-being   41

    5.1   Blending Eastern and Western Ideals   41
    5.2   Sustainable Happiness and Well-Being   42
          5.2.1   Income Inequality   42
          5.2.2   Relationships, Health, and Meaning   44
    5.3   Long-term/short-term balance   46
    References   47

6   The Bubble Bursts: Population and Pollution
    Become Our Concern   49

    6.1   Population Growth   50
          6.1.1   Population Statistics   51
          6.1.2   Fertility Rates   51
          6.1.3   Birth and Death Rates   53
          6.1.4   Median Age, the Elderly Dependency Ratio,
                  and the Economy   54
    6.2   Increased Pollution   56
          6.2.1   Greenhouse Gas   57
          6.2.2   Air Pollution   61
          6.2.3   Municipal Solid Waste   61
          6.2.4   Radioactive Waste   63
          6.2.5   Industrial, Agricultural, and Human Wastes   65
    6.3   Interaction Among Population, Economy, and Environment 66
    References   70

7   Applying the Brakes: Factors That Limit Growth   73

    7.1   Ancient Civilization and Limits to Growth   73
    7.2   Carrying Capacity   76
          7.2.1   Determinants of Carrying Capacity   76
          7.2.2   Carrying Capacity and Our Mental Model   77

7.3   Limiting Factors   77
        7.3.1   Energy Supply   78
        7.3.2   Water Supply   82
        7.3.3   Food Supply   85
        7.3.4   Forests   88
        References   90

8   The "I"s Have It: A Systems View of Sustainability   95

8.1   Integration and Interdependence   95
8.2   The system diagram   96
        8.2.1   Step 1: Integrating Economy and Environment   97
        8.2.2   Step 2: Integrating Society with Economy
                and Environment   100
8.3   Mental Model and Integrated System Compared   106
8.4   A Video of the Future   106
        8.4.1   Future 1: Population Increases   107
        8.4.2   Future 2: Population Decreases   108
8.5   Imperfections and Lessons   109
        8.5.1   Short Term versus Long Term   109
        8.5.2   Self-Interest versus Community Interest   110
        8.5.3   Economy, Environment, and Society   110
        References   111

9   Creating Balance: Effective Interventions   113

9.1   Sustainability Solutions: Systemic or Suboptimal?   114
        9.1.1   Leverage   114
        9.1.2   Generic Types of Intervention   115
9.2   Analytic Approach   115
9.3   Areas of Intervention   118
9.4   Proposed Actions   121
        9.4.1   Three Levels of Effectiveness   121
        9.4.2   Opposing Effects   121
        References   124

10   Pieces of the Puzzle Level I: Paradigm Shifts   125

10.1   Synergistic Plan to Achieve Sustainability   125
10.2   Building the Foundation   126
10.3   Area 1: Mental Model   126
        10.3.1 Arouse Moral Commitment Through Fear and Inspiration 128
        10.3.2 Disseminate Information in Multiple Formats and Media 128
        10.3.3 Ongoing Efforts: Research and Information Dissemination 129
10.4   Area 2: Balanced Cultural Values   129
        10.4.1 Champion Balanced Cultural Beliefs for Sustainable
                Happiness and Well-Being   131
        10.4.2 Broaden Society's Metrics for Success   132
        10.4.3 Ongoing Efforts: New Economics and New Metrics   132
        References   133

11   Pieces of the Puzzle Level II: Structural Changes   135

11.1   Altering Feedback Loops   135
11.2   Area 3: Energy Cost   137
    11.2.1   Recover Full Cost of Energy Products and Services   137
    11.2.2   Prepare for Cost of Living Increases   137
    11.2.3   Ongoing Efforts: Attempts to Raise Energy Costs   138
11.3   Area 4: Births Per Year   139
    11.3.1   Educate and Indoctrinate   139
    11.3.2   Unite National Policies   140
    11.3.3   Increase Education Levels   140
    11.3.4   Ongoing Efforts: Policies for Population Control
        and Population Growth   140
11.4   Area 5: Median Age   142
    11.4.1   Implement Retirement-Related Changes   143
    11.4.2   Balance Median Age Among Nations   143
    11.4.3   Ongoing Efforts: Retirement Laws, Immigration,
        Retraining, and Incentives   144
References   145

12   Pieces of the Puzzle Level III: Transition to the Future   147

12.1   Attending to the Present   148
12.2   Long-Term Versus Short-Term Conflict Resolution   148
12.3   Area 6: Environmental Technology and Policies   150
    12.3.1   Reduce Pollution   150
    12.3.2   Repair the Environment   153
    12.3.3   Ongoing Efforts: Reducing GHG, Reforesting,
        and Climate Engineering   154
12.4   Area 7: Energy Technology and Policies   156
    12.4.1   Increase Investment in Renewable Energy Technologies   157
    12.4.2   Conserve Energy Resources   157
    12.4.3   Ongoing Efforts: Green Energy and Energy Conservation   157
12.5   Area 8: Food and Water Technology and Policies   158
    12.5.1   Increase Food Production   160
    12.5.2   Conserve Water   160
    12.5.3   Develop New Water Sources   161
    12.5.4   Ongoing Efforts: Sustainable Agriculture and
        Water Conservation   161
References   162

13   From Bud to Blossom: Nurturing Sustainable Stewardship   167

13.1   From Anxiety to Hope   168
13.2   From Mental Model to Integrated System   168
    13.2.1   Economic Roots Are Deep   168
    13.2.2   Deriving Relationships from Current Trends   169
    13.2.3   A Three-Level Framework for Intervention   170

13.3 From Integrated System to Sustainable Stewardship 170
    13.3.1 The Collective 171
    13.3.2 The Individual 176
    References 179

14 The Global Commons and the Uncommon Globe: System Insights and Conclusions 181

14.1 The Global Commons 182
    14.1.1 Tragedy of the Commons and Sustainability 184
    14.1.2 Repercussions of Mismanaging the Global Commons 185
14.2 Insights from Systems Thinking 185
    14.2.1 Inertia, Balance, and Perspective 186
    14.2.2 Systems Thinking and Effective Intervention 188
14.3 Where to Now? 188
    References 189

Glossary 191
Index 195

Preface

"Two steps ahead on the climate; California might be able to cash in on EPA rule"
"Spending to fight drought gets little support"
"The Asian nation's booming dairy industry is hungry for California alfalfa"

In less than the span of a week, these headlines appeared in the *Los Angeles Times*.[1] Every day, some event, fact, or opinion relating to sustainability pops up. The goal of sustainability – to meet our needs today without diminishing the ability of future generations to meet their needs – is so vital that it is swiftly becoming common fare. Although piecemeal information from the media may cause us to shake our heads in dismay for a minute or two, we soon go on about our day and leave anxiety far behind.

Sustainability, the subject of this book, has become an obsession for me. It is hard to listen to the latest scoop about changing conditions on Earth without wanting to do something about it. Although technologies that put the first man on the moon and created the Internet have dramatically transformed our lives over the past 50 years, this transformation pales in comparison to the transitions yet to come in the next half century.

Being constantly bombarded with news about population escalation, peak oil, fracking, droughts, hurricanes, economic downturns, or global warming is like anticipating the "big one" – the earthquake that will open the ground beneath us. The convergence of many sobering conditions over the next decades will test our fundamental views and necessarily focus our attention on sustaining society. We already see hints that our current fixation on economic growth must lose its intensity, that quality of life will take on new meaning, and that fragmented problem solving will no longer work. Thus, in addition to altering our individual ways of life, we must transcend our current mindsets about the world.

Now we come to the purpose of this book. I wrote it to join with others who have ceased handwringing and are acting so that future generations can flourish. Because my passion is to view complex problems holistically, I wanted to combine ragtag events and trends into a picture of dynamic relationships that describe sustainability's challenges. Using a discipline called "systems thinking" makes it easier to visualize what is happening and allows us to investigate possible solutions.

The system defined by these trends does not include everything that contributes to or detracts from sustainability, for that would involve a miracle of

---

[1] From *The Los Angeles Times*, June 3, June 6, June 9, 2014.

perception and comprehension. It does, however, incorporate the foremost factors and their interactions. These primary elements come together as a systems diagram that forms the framework for many sustainability-related issues.

This book is intended for a diverse audience who seeks answers, wants to understand, and desires to help. Those who study sustainability will gain insights on issues and on unintended consequences of actions. Aficionados of economics will appreciate how intricately the economy is entwined with our lives and with the environment, and may be surprised by the pervasive effects of economic growth. Local, national, and international policymakers can gain new perspectives on what prevents and what encourages sustainability. Organizations and individuals can appreciate how to live well today and preserve the quality of life for tomorrow. Systems thinkers will welcome the application of systems theory to this multifaceted conundrum.

* * * * * * * *

My hope is that the book will open our eyes and our hearts to view life in the future as part of our responsibility today. We have not taken this responsibility seriously; we have neither considered the bigger picture in our behaviors nor worried about our legacy. We are, in fact, leaving younger generations to mop up after us. My desire, then, is to inform and inspire each of you to move from *me* to *we*; to join together and become stewards for tomorrow's world.

Before we launch into sustainability's sticky issues, I want to thank those who rallied around my excitement for the book and shared their concerns about the world. Kathleen Pagel read and reread, edited, visualized concepts, provided reference material, and encouraged me at every step. Susan Smith's thoughtful alternative views improved the book's focus and content. Dr. Susan Rogers energized me with her intuitive insights about living as part of a greater whole. Dr. Randy Hodson's far-reaching expertise provided an exceptional sounding board about the effects of population dynamics on the economy. Sherri Scofield added depth to the reference material and to the scope of concerns.

My publisher, Elsevier, is a prominent model of professionalism and integrity. Senior Acquisition Editor, Dr. Scott Bentley, provided substantive, constructive, and thoughtful suggestions; his enthusiasm was heartening. The help, patience, and efficiency of Mckenna Bailey, Editorial Project Manager and Melissa Read, Project Manager from ReadIt Publishing smoothed the transformation of draft to finished product. Their encouragement and publishing expertise were outstanding. Advice from Cindy Minor, Portfolio Marketing Manager, was especially practical and insightful. All these folks make an awesome team!

I found that hope is well placed with future generations. Four young ladies – Vicki Ensign, Beth Ensign, Mandy Limond, and Sydney Jarboe – shared their thoughts on the characteristics, goals, and concerns of the millennial generation. Their high energy, optimism, and world-minded perspectives influenced my ideas about the road ahead.

And of course, my husband Tim patiently listened to my theories, collected relevant information, and edited illustrations. His penetrating questions, contrasting views, and insights crystallized ideas and clarified points.

Finally, I appreciate those forward thinkers whose work on sustainability, environment, and systems thinking is interwoven throughout the book. Though I may never know you, I thank you: Donella Meadows (who was taken too early from this world), John Sterman, Richard Heinberg, Jared Diamond, Jeremy Rifkin, Lester Brown, Harland Cleveland, Andres Edwards, Fritjof Capra, and so many others. Your dedication and courage to expose "what is" from your unique perspectives are admirable.

Thanks to you all for helping to bring this vital message to a broad audience. Now, let us view the world through the eyes of those who will confront sustainability full in the face. Please join me in a search for solutions.

Karen L Higgins

About the Author

KAREN L. HIGGINS is an adjunct professor of management at the Drucker/Ito Graduate School of Management, Claremont Graduate University in California where she teaches Project Management, Ethical Leadership and Systems Thinking. As president of *Élan Leadership Concepts* and in her association with *Effective Edge* in Austin, Texas, she provides training in leadership and organizational culture. During her previous career with the Naval Air Warfare Center, Weapons Division at China Lake, California, she held various engineering, line management, and project management positions. While at the Weapons Division, Dr. Higgins became a member of the Navy's Senior Executive Service and held the top civilian position of Executive Director and Director for Research and Engineering. She received a BS in mathematics and an MS in electrical engineering from the University of Idaho, and an MBA and PhD in executive management from Claremont Graduate University. She lives with her husband in southern California, near her children and grandchildren.

Introduction – the Looming Challenge of Sustainability

Failures to manage the environment and to sustain development threaten to overwhelm all countries. ...These problems cannot be treated separately by fragmented institutions and policies. They are linked in a complex system of cause and effect.

World Commission on Environment and Development, 1991

Humankind is a miraculous paradox. On one hand, we are compassionate, worry about the future of our children, and reach beyond our earthly bounds to explore outer space. On the other hand, we are selfish, focus on the here and now, and argue about the neighbor's garbage that litters our gutters. We can dispassionately hear about remote events that are concerning, but that do not concern us; hurricanes, fires, power outages, drought, blizzards, heat waves, smog, civil uprisings, wars, and even global warming are others' problems unless we are in their midst. We know that many of our actions, such as eating too much fat or spending money that we do not have on things that we do not need, are not good in the long run, but somehow we cannot help ourselves.

So, we wonder, with all these inconsistencies: What will our society look like in a hundred years? Is it possible for our dual-natured species to meet today's needs without affecting the ability of future generations to meet their own needs?[1] Can the present that we know and feel and see compete with a future that we will never witness? In other words, is the *sustainability* of our current way of life on the planet possible?

While these questions may be the fanciful grist of science fiction writers, they should be a serious matter for citizens of the world. Yet, they lurk in the underbrush of our awareness, often disregarded in favor of superficial and immediate concerns about what to have for dinner or when to buy gas. We must therefore force the issue of sustainability into the sunlight. How else can we look into the innocent faces of our children and grandchildren? So, let us make it personal. Say the names of children you know and think of the children they may have. What are you doing to secure their future on Earth?

Today's society faces unprecedented challenges that will become millstones around the necks of our progeny unless we take serious and rapid action. Philosophically it is curious that we do not take better care of our planet. Why are there bold words, but only a few actions around the fringes of the problem? My conclusion is twofold.

---

[1]Definition of sustainable development by the World Commission on Environment and Development,1991.

First, I believe it is difficult to view the whole of sustainability in a way that allows us to internalize it. It is not easy to see how global trends that affect society, economy, and environment (the three pillars of sustainability) are part of the same picture. We must ask many questions to merge these trends in our minds. For example: How do pollution and carbon dioxide relate to the economy or affect happiness? How does our addiction to material goods relate to the environment? What role does population growth play in the economy?

It is even trickier to envision exactly how individuals fit into this picture. Although our brains have a capacity greater than other creatures on Earth, they still function within limits. Old but familiar research places our short-term mental capacity for remembering lists at seven plus or minus two items – not enough to wrap our minds around the many issues of sustainability at the same time (Miller, 1956). Regardless of whether this number is exact, it causes us to acknowledge that as individuals, we cannot readily grasp highly complicated subjects; we tend to look at complex problems either a piece at a time or tied to something familiar.

Second, we have a mental protective mechanism that causes us to strive for self-preservation. Whether humans are by nature egoistic to ensure their survival or altruistic to flourish within collectives has been debated for centuries. Short-term egoism, as well-known psychologist B.F. Skinner argues, is "exactly the psychological trait that makes modern humans prone to environmental tragedies of the commons" (Gardner and Stern, 1996). In other words, when it comes to making immediate sacrifices to preserve a future we will never see, particularly when these sacrifices involve common resources such as energy, clean air, or water, our actions favor short-term personal gratification and neglect long-term consequences to the well-being of society. No doubt most of us exhibit a mix of both, depending on our circumstances; but a balance between present and future matters when it comes to sustainability.

By tackling sustainability in an unconventional way, this book deviates from the surfeit of books, articles, reports, and research on the topic. It addresses the two reasons that make it difficult to understand sustainability: (1) the limits of our processing power; and (2) our predisposition toward short-term thinking. Like many other books and articles on the subject, this book presents statistics for major world trends and discusses their significance. More importantly, however, rather than dealing with these trends one by one, it *integrates* them into a system of interdependencies. To overcome our struggle with complexity, it uses the visual language of systems thinking first to define the preponderant mental model[2] that guides day-to-day behaviors and shapes world trends, and then to create an integrated system diagram that portrays the fundamentals of how the world actually works, particularly with regard to sustainability.

Viewing sustainability from this holistic perspective allows us to grasp the profound predicament we face when sustainability and limited resources butt

---

[2]Mental models reflect our internal beliefs about how the world works.

heads with a way of life that promotes abundance and short-term thinking. By comparing our mental model with the integrated system diagram, we find that current lifestyles, with their focus on the present and their disproportionate emphasis on the economy, have disrupted our ability to achieve sustainability. Existing only within the confines of our narrow perceptions – the bubble that constricts our actions – can bring unintended and harmful consequences in the future.

A caveat to this system approach is in order here. If we were to incorporate every possible issue that could contribute to a good life in the future, we would be mired in the mud of too much complexity. Thus, as with any system of interest, we have defined the boundaries of the sustainability system and have incorporated its most basic concerns such as environmental damage, population dynamics, availability of food and water, and dependence on economic growth. Other factors, such as biodiversity, urbanization, and infectious disease (which are well-described by Brown in *Beyond Malthus*) (Brown et al., 2000), and societal elements such as discrimination, religion, and politics are not explicitly discussed. However, their interaction can be placed in the context of the system described in this book.

Finally, the book weaves in the human element, noting how the dualistic nature of humans, who must reconcile present with future and self with community, may be at odds with sustainability's goals. Rather than abandoning all hope, it then proposes a comprehensive set of actions to recalibrate our mental model, burst our myopic bubble, and aid our pursuit of sustainability. It encourages us to see beyond our own backyards and our own lifetimes.

By the book's end, we will realize that sustainability is the ultimate state of dynamic balance in which competing concerns are blended in a harmonious dance of flow and change. Thus, achieving sustainability requires a balance among society, economy, and environment, between self and others, and between present and future. We will understand that the system we are a part of has such massive inertia that it will take decades – perhaps even a century – to change, *but* that change is both possible and imperative. Such change requires courage and lifestyles that embrace stewardship; it requires collaboration among groups and profound alteration of individual behaviors.

REFERENCES

Brown, L., Gardner, G., Halweil, B., 2000. Beyond Malthus: Nineteen Dimensions of the Population Challenge. Earthscan, London.

Gardner, G., Stern, P., 1996. Environmental Problems and Human Behavior. Allyn and Bacon, Boston.

Miller, G., 1956. The magical number seven, plus or minus two: some limits on our capacity for processing information. Psychol Rev 63(2), 81–97.

World Commission on Environment and Development, 1991. Our Common Future. Oxford University Press, Oxford.

Chapter 1

The Secret's in the Overlap: Sustainability as an Integrated System

We are born from the earth, supported by the earth, and return to the earth, and we need the earth to live.

—Michael Stone (Stone, 2009)

By taking a comprehensive look at the interconnections among ecological, economic and equity issues ...we are more likely to seek and implement lasting solutions.

—Andres Edwards (Edwards, 2005)

The notion of sustainability has a long history. The Old Kingdom of ancient Egypt strived to take its civilization into eternity through customs, culture, religion, and irrigation technology. Millennia later and nearly halfway around the world, the Mayans of southern Mexico practiced ecological engineering to conserve water and food, and to preserve a way of life for their descendants. The sophistication of these future-thinking civilizations is apparent from their once vibrant physical remains (see Fig. 1.1).

In more recent times, naturalists Ralph Waldo Emerson and Henry David Thoreau in the 1800s and John Muir in the early 1900s championed a close relationship with Earth. Their writings recognized that harmony between man and nature was possible. About 90 years ago, Russian biogeologist Vladimir Vernadsky was one of the first modern-day scientists to acknowledge the implications of this interdependence. His eureka that Earth is a self-contained sphere which supports life and that "life exists only in the biosphere" (Vernadsky, 1998) was soon followed by other researchers' "big" ideas.

Prescient in her awareness of this interdependence, marine biologist Rachel Carson made the relationship between humans and Earth more concrete. She identified the destructive effects of the man-made pesticide DDT on the world's food supply. At the time, the chemical industry denounced her 1962 book *Silent Spring*, asserting that its findings would hurt hundreds of thousands of people if farmers could not use DDT. Even so, she changed our perception of man's

K.L. Higgins: Economic Growth and Sustainability. http://dx.doi.org/10.1016/B978-0-12-802204-7.00001-3

1

FIGURE 1.1 Mayan ruins (left) and Egyptian ruins (right). *Source: Palenque, Mexico by Ricraider (2012), retrieved from <http://en.wikipedia.org/wiki/File:The_Palenque_Palace_Aqueduct.jpg>; Sphinx and pyramid in Cairo by Przemyslaw Idzkiewicz (2005), retrieved from <http://en.wikipedia. org/wiki/File:Cairo,_Gizeh,_Sphinx_and_Pyramid_of_Khufu,_Egypt,_Oct_2004.jpg>.*

relationship with the environment. These forays into a new way of thinking set the stage for greater consciousness of life's interdependence with Earth.

1.1 ENVIRONMENTALISM AND SUSTAINABILITY

In the mid-nineteenth century – the same century that these philosophers and scientists connected the consequences of man's actions on Earth – a social movement called *"Environmentalism"* emerged. This movement evolved from the vital need to decrease the ruinous air pollution and chemical wastes produced by technological advances of the Industrial Revolution. These early environmental concerns were a forerunner of the "Sustainability Revolution" that we know today (Edwards, 2005).

1.1.1 The Sustainability Revolution

Over the past decades, the term "sustainability" has almost become a household word like "kleenex" or "bandaids" that began as brand names and ended up as the generic description of a product. Ecologically speaking, sustainability refers to how organisms remain diverse and adaptive; it means that they are able to endure over time. In today's vernacular, however, sustainability often takes on the simpler definition of what has been called "sustainable development," that is, the ability to meet "the needs of the present without compromising the ability of future generations to meet their own needs."[1] In this book, we refer to this latter definition when we talk about sustainability.

---

1. Definition of sustainable development from World Commission on Environment and Development (1991).

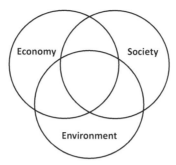

FIGURE 1.2   Overlapping circles of sustainability. *Source: Adapted from overlapping circles of sustainable development defined at The World Conservation Congress in Bangkok, Thailand, November 17–25, 2004 (IUCN, 2004). Reproduced with permission of the International Union for Conservation of Nature.*

The first *Earth Day* in 1970 marked our entrée into contemporary sustainability with formal discussions that united environment and economy. International conferences, new regulations, and organizations such as the Environmental Protection Agency appeared in the 1970s and the 1980s. Concerns escalated into an overarching goal to create a high-quality environment and healthy economy for all people. In 1969, the International Union for Conservation of Nature extended the bounds of environmentalism with a mandate to achieve the "highest sustainability of life on our world" (Adams, 2006). World leaders formed the Brundtland Commission in 1983 to focus on sustainability issues.[2] They introduced the concept of three overlapping areas – economy, environment, and society – to describe sustainability (see Fig. 1.2). The secret to sustainability, they suggest, is to maintain balance, that is, to give these areas equal weight in our decisions and actions.

In addition to studies and policies, quantitative analyses were popularized in the late 1960s. An informal group of businessmen, government officials, economists, and scientists founded the Club of Rome in 1968 to explore their shared concern about humanity. Using historic trends and computer models, the group published disturbing scenarios for population growth, pollution, and depletion of natural resources.[3] Escalation in these areas, they predicted, could cause collapse of physical growth on Earth and reduce our quality of life sometime in the twenty-first century. Backlash at the time countered that "there are no effective limits" and that "technological change can substitute for social change" (Meadows, 2012).

Around the same period, Alvin Toffler shocked the world with his claim that by exploiting "technology for immediate economic advantage, we have turned our environment into a physical and social tinderbox" (Toffler, 1971). At the

2. The Brundtland Report, also called "*Our Common Future*", was first published in 1987 after the group was dissolved (see World Commission on Environment and Development, 1991).
3. Findings were published in the popular yet highly criticized book *The Limits to Growth*, Meadows et al. (1972).

time, few believed we would ever deplete our natural resources or that technology had a dark side.

Thus, the seeds of sustainability that were planted by a few forward thinkers in the nineteenth century took root in the 1970s and are maturing today. Recent attention recognizes that our growth-dependent lifestyles are radically reshaping our climate, our economy, and our future well-being.[4]

1.1.2 Sustainability and Interdependence

Sustainability's three overlapping circles illustrate that our ability to sustain life on Earth is an interdependent, highly complex issue. These domains encompass our every action and decision, primarily because our lives are inextricably entwined within them. The products we buy have been touched in some way by someone in some distant place. Business, social, and religious groups unite goals and ideologies around the globe. Jobs in one country depend upon the social and economic welfare of those in other countries who buy imported goods and services. Local pollution from coal or automobiles or nuclear meltdown affects the rest of the world. For example, pollution originating from China's industrial plants reaches the West Coast of the United States in about 6 days; their export-related pollution contributes as much as 24% of sulfate concentration in the western United States (Lin et al., 2014). In addition to these physical interdependencies, issues of sustainability have no temporal bounds; past and current actions profoundly affect the future.

To understand such an intricate subject and to compensate for our human limitations, we need a way to visualize and simplify sustainability's dynamic interdependencies; we need a way to investigate what I call our "*integrated system*" and to identify its challenges. One such approach is well suited to the job: systems thinking.

1.2 SYSTEMS THINKING

Systems thinking is an integrative way to view a large and complex issue as part of a group or system of elements that function together as a whole. With a multidisciplinary heritage that includes philosophy, sociology, biology, and engineering, systems thinking draws from its quantitative cousin, system dynamics. System dynamics relies on diagrams of influence, that is, constructs called "*causal loop diagrams,*" to represent relationships and interactions among elements. It then applies engineering equations to these diagrams and develops computer models of system behavior. Likewise, systems thinking creates *causal loop diagrams* but uses interdependencies and trends rather than specific data or equations to describe a system's behavior.[5]

---

4. See, for example, Gore (2013), Heinberg (2011), and Smith (2010).
5. For more detailed descriptions of causal loop diagrams, see Sterman (2000) and Higgins (2013).

The beauty of systems thinking is that one can gain deep insight from the big picture perspective inherent in these diagrams. Causal loop diagrams allow us to see patterns and root causes which are the basis for long-term systemic solutions. Solutions to knotty issues do not come from putting Band-Aids on symptoms. They come from finding origins and areas in which small actions have large effects. Well-known systems thinking guru Peter Senge describes this "bottom line of systems thinking" as *leverage*, or "seeing where actions and changes in structures can lead to significant, enduring improvements" (Senge, 1990). We talk more about levers and leverage in Chapter 9 when we identify actions that support sustainability.

Simplistically, systems thinking can be thought of as a way to depict a set of events or phenomena using an assortment of interlocking gears like those in the child's toy in Fig. 1.3. Rather than examining each gear individually to follow its clockwise- or counter-clockwise rotation or to count its teeth, systems thinking considers how the movement of one gear affects the next and the next until it recognizes that all gears are intertwined and spinning – even those that are far apart. Keep these spinning gears in mind as we explore the causal loop diagrams and constructs of systems thinking in the following sections.

1.2.1   Systems Thinking Constructs

Different from other types of analyses, systems thinking relies on the notion of feedback to understand how something operates. Rather than viewing an issue as a linear chain of cause and effect with some outcome at a given *point in*

FIGURE 1.3   Interlocking gears.

time, systems thinking considers the conditions under which an effect can loop or feed back to influence the original cause. It characterizes the human, physical, and institutional interactions that shape behavior *as time passes*. Feedback loops reveal how human behavior changes our world and how these changes return to alter that behavior (Sterman, 2013).

1.2.1.1 Feedback Loops and Delays

There are two basic types of feedback loops: (1) reinforcing and (2) balancing. Reinforcing feedback generates growth or decay; behavior builds on itself and drives a system the way it is already going. *"Vicious and virtuous circles"*, *"death spirals"*, and *"escalation"* are common terms for reinforcing feedback. Reinforcing feedback is apparent in an undisturbed savings account that earns interest; earned interest then increases the balance, which increases the interest earned, which further increases the balance – and so on. Untreated cancer is another case of reinforcing feedback. When cancer cells multiply, they create more cancer cells which, unless somehow thwarted, also multiply until the cancer has spread and the body is overwhelmed.

Like steering a car, balancing feedback stabilizes; it pushes behavior toward a given state or goal and works to maintain that condition. Phrases such as *"meet your goals,"* *"hold it steady,"* or even *"work within the rules"* and acts such as setting the thermostat in your home exemplify balancing feedback. Balancing feedback also constrains a system's behavior to stay within the bounds of its capacity, that is, its limits. As a simple example of a limit, consider how many times you can fold a paper napkin in half and half again. After a certain number of halvings (around six), the folded napkin becomes either too thick or too small to fold again. It has reached its *fold* limit. In a different context, cultural norms also operate as balancing feedback. In this case, human behavior can be guided or restricted to fall within the belief systems of countries, groups, or organizations.

Inherent in feedback loops are time delays or lags between cause and consequence. Delays are a crucial aspect of systems thinking because they separate cause and consequence; it can be difficult to tell which action had what effect. Delays are often a system's nemesis; they can produce unintended consequences. During the 2008 financial crisis, for example, although relaxed credit requirements for home loans were initially beneficial, they had disastrous consequences years later.[6] Later chapters describe examples of how delay affects sustainability.

1.2.1.2 Causal Loop Diagrams

Causal loop diagrams (CLD) are the visual language of systems thinking. They depict feedback as a loop built from arrows of influence between one element and

---

6. See Higgins (2013) for a systems analysis of this crisis.

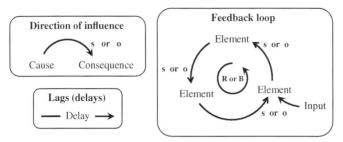

FIGURE 1.4   Components of a causal loop diagram. *Source: Adapted from Higgins (2013). Reproduced with permission of Elsevier.*

another, that is, between cause and consequence. In the loop, each element is both cause and consequence. The arrowhead indicates the direction of influence. It is labeled with an "s" (same) when cause and consequence move in the same direction (up or down) or an "o" (opposite) when consequence moves opposite of cause. An "R" or "B" in the center of a loop labels it as reinforcing or balancing. In this book, we simply name the loop to clarify its function. Time delays are annotated as *delay* on an arrow of influence. Fig. 1.4 shows the components of *causal loop diagrams.*

Most systems have multiple feedback loops that operate interdependently. A larger system diagram incorporates these relationships to characterize the behavior of the entire system. System diagrams provide both panoramic and dynamic perspectives.

1.2.2   Boundaries and Limits

Two other important aspects of systems thinking that apply to sustainability are the *system boundary* and a commonly used construct called *"limits to growth."* Both represent system constraints.

1.2.2.1   System Boundaries

Because systems can be of any size – from atoms to humans to organizations to the universe – we must restrict our field of interest lest the system is too small to give an accurate picture or too large to comprehend. A *system boundary* delineates what is inside and what is outside our immediate concern. While external elements may influence the system or may be an output from it, they are not considered an integral part and thus are treated as remote phenomena. System diagrams in this book denote this boundary with dashed lines; external elements – inputs and outputs – lie outside these lines.

1.2.2.2   Limits-to-Growth

Natural systems such as population or the economy cannot grow beyond certain bounds. The *limits-to-growth* construct describes how a system behaves

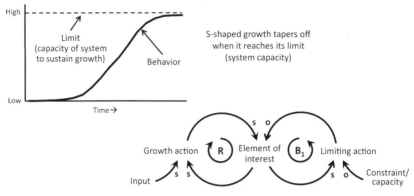

FIGURE 1.5 S-shaped growth. *Source: Adapted from Higgins (2013). Reproduced with permission of Elsevier.*

when it hits these bounds, that is, when it reaches or exceeds its capacity to sustain growth.[7] There are two fundamental reactions to this condition: *s-shaped growth* and *overshoot-and-collapse*. For s-shaped growth, elements in the system grow quickly at first and then slow down until the system gradually reaches its capacity. For overshoot-and-collapse, elements grow uncontrollably, overshoot their capacity and then decay. This situation generally occurs in the presence of delayed understanding or response, when it is too late to contain the growth smoothly.

1.2.2.2.1 S-Shaped Growth

In Fig. 1.5 s-shaped growth is a reinforcing loop connected to a balancing loop that constrains growth. As the balancing loop gradually grows stronger (i.e., the system nears its limits), growth tapers off and the system remains just below its capacity. Its behavior resembles a lazy "S."

1.2.2.2.2 Overshoot-and-Collapse

Fig. 1.6 presents the case of overshoot-and-collapse that combines a reinforcing loop of growth with two balancing loops that try to bring the system toward a goal or equilibrium state. At least one of these balancing loops eventually overpowers the reinforcing loop after it has already exceeded its capacity. Once this condition occurs, growth turns to decay and the system's ability to grow in the future is damaged. Overshoot-and-collapse exemplifies a harmful aspect of delay in a system. The figure shows how delays in balancing loops prevent an immediate stop or gradual slowing; growth continues until it is too late. In other words, actions that take too long or are not applied early enough may not work in time to contain runaway behavior.

---

7. For more details on the limits-to-growth archetype, see Senge (1990) and Higgins (2013).

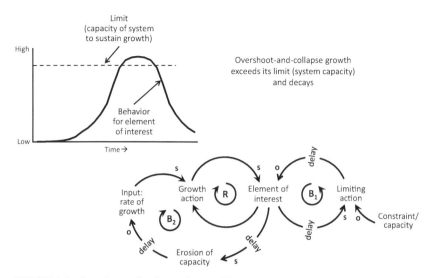

FIGURE 1.6 Overshoot-and-collapse. *Source: The graph at the top is adapted from Higgins (2013). Reproduced with permission of Elsevier, Inc. The CLD at the bottom is adapted from Sterman (2000). Business Dynamics: Systems Thinking and Modeling for a Complex World, Irwin McGraw-Hill, New York. Reproduced with permission of The McGraw-Hill Companies.*

We will see in later chapters that overshoot-and-collapse is a prevalent reaction to some attempts at sustainability. We will also recognize that how we set system boundaries makes all the difference in what we see as outcomes.

1.2.3   Systems Thinking Applied

Applying a systems approach to sustainability is not new. Its rich history began in the late 1960s with ecologists' study of the relationships between living organisms and their environments. From the 1970s through the 1990s, notables like Dana Meadows (see Meadows et al., 1972) and Jay Forrester relied on system dynamics to research the capacity of Earth's natural systems to sustain life. More recently, in recognizing that "our society is unsustainable and getting worse fast," systems expert John Sterman suggests that we often ignore "the networks of feedback that bind us to one another and to nature." He proposes that systems thinking can help us understand how "the world operates as a system" and how to meet the challenges posed by this earthly system (Sterman, 2013).

In considering sustainability as an integrated system, we must first acknowledge that a system does not *choose* to be good or bad; it simply adapts to its environment and behaves in ways determined by the interactions among its parts. Early systems thinkers understood this principle. Stafford Beer, for example, noted that "society is a dynamic system" whose outputs are *not* "accidental mistakes" but come from how that system operates (Beer, 1993). Thus, believing that a system will automatically avoid damage or ensure human survival is naïve at best. Furthermore, because systems thinking is inclusive and comprehensive, it has no "side effects – just effects" (Sterman, 2013). Thus, to change a system's behavior, we must change the relationships and delays within it. This principle is a mainstay for solving sustainability challenges.

1.3 LESSONS FOR THE FUTURE

At the beginning of the chapter, we referred to the sustainability consciousness of two ancient civilizations and implied that contemporary civilizations can learn from these holistic perspectives. In addition to this advice, we can elicit a warning from the not-so-positive demise of both. The ancient Mayans, who in spite of their best efforts and their reverence of Earth, lacked the technology and the ability to adapt to changes in the environment; their civilization perished from a prolonged drought in the ninth century AD (Spurgeon, 2000). NASA suggests that they may have had a hand in this drought; by clearing forests to feed their expanding empire, they could have altered their regional climate (Brown, 2011).

The Egyptian Old Kingdom collapsed around 2200 BCE after a 50-year drought. Although strong leadership and stronger armies resurrected this civilization, it took centuries to evolve into one that was successful, albeit vastly different from earlier times. This new society flourished until it came under Roman rule in the fourth century BCE and finally declined after the Arab conquest in the seventh century AD. Its 3000-year existence epitomizes success of a sustainable civilization. Some even claim that it was far more advanced than previously believed and suggest that the precise layout of their pyramids signifies their use as power generators (Dunn, 1998). But, in spite of its powerful technologies and vigorous social structures, this ancient civilization eventually deteriorated just like others that followed.

Even in these instances of decay there are lessons to be learned. We can set aside the arrogance that makes us believe nothing will happen to us. Without becoming the storybook character Chicken Little who warned that the "sky is falling" when an acorn fell on his head, we can open our minds to the bigger picture and adapt to conditions that threaten current well-being and the well-being of future generations – the happiness and security of our grandchildren's grandchildren.

And now, with the help of systems thinking and through this lens of optimism, we examine economic, social, and environmental trends that define our integrated system and shape the future. The next chapter describes our mental model – our *beliefs* about how the world works. In subsequent chapters, we propose how the world – this integrated system – *actually* works with respect to sustainability. We also identify points at which change can dampen destructive consequences.

REFERENCES

Adams, W., 2006. The future of sustainability: re-thinking environment and development in the twenty-first century. Report of the IUCN Renowned Thinkers Meeting, January 29–31, 2006. The World Conservation Union. Retrieved from <http://cmsdata.iucn.org/downloads/iucn_future_of_sustanability.pdf>.

Beer, S., 1993. Designing Freedom. House of Anansi Press, Toronto, ON.

Brown, L., 2011. World on the Edge: How to Prevent Environmental and Economic Collapse. W. W. Norton, New York.

Dunn, C., 1998. The Giza Power Plant: Technologies of Ancient Egypt. Bear & Company, Rochester, VT.

Edwards, A., 2005. The Sustainability Revolution: Portrait of a Paradigm Shift. New Society Publishers, Gabriola Island, BC.

Gore, A., 2013. The Future: Six Drivers of Global Change. Random House, New York.

Heinberg, R., 2011. The End of Growth: Adapting to Our New Economic Reality. New Society Publishers, Gabriola Island, BC.

Higgins, K., 2013. Financial Whirlpools: A Systems Story of the Great Global Recession. Elsevier, Oxford.

IUCN, 2004. The IUCN Programme 2005–2008: Many Voices, One Earth. Adopted at The World Conservation Congress, Bangkok, Thailand, November 17–25, 2004. Retrieved from <http://cmsdata.iucn.org/downloads/programme_english.pdf>.

Lin, J., Pan, D., Davis, S., Zhang, Q., He, K., Wang, C., Streets, D., Wuebbles, D., Guan, D., 2014. China's international trade and air pollution in the United States. PNAS 111 (5), 1736–1741, Retrieved from <http://www.pnas.org/content/early/2014/01/16/1312860111>.

Meadows, D.L., 2012. Is it too late for sustainable development? Paper presented at the Smithsonian Consortia Symposium, Perspectives on Limits to Growth: Challenges to Building a Sustainable Planet. Retrieved from <http://www.smithsonianmag.com/science-nature/is-it-too-late-for-sustainable-development-125411410/>.

Meadows, D.H., Meadows, D.L., Randers, J., Behrens, III, W., 1972. The Limits to Growth: A Report for the Club of Rome's Project on the Predicament of Mankind. Universe Books, New York.

Senge, P., 1990. The Fifth Discipline: The Art & Practice of the Learning Organization. Doubleday/Currency, New York.

Smith, L., 2010. The World in 2050: Four Forces Shaping Civilization's Northern Future. Dutton, New York.

Spurgeon, C., 2000. Mayan archaeology: Ancient Mayan Civilization. Retrieved from <http://mayanarchaeology.tripod.com/id2.html>.

Sterman, J., 2000. Business Dynamics: Systems Thinking and Modeling for a Complex World. Irwin McGraw-Hill, New York.

Sterman, J., 2013. 5 ways 'systems thinking' can jumpstart action. Network for Business Sustainability. Retrieved from <http://www.greenbiz.com/blog/2013/12/09/systems-thinking-climate-change-lessons-action>.

Stone, M., 2009. Yoga for a World out of Balance: Teachings on Ethics and Social Actions. Shambhala, Boston.

Toffler, A., 1971. Future Shock. Bantam, New York.

Vernadsky, V., 1998/1926. The Biosphere. In: Langmuir, D. (Ed.), Complete Annotated Edition. Springer Verlag, New York.

World Commission on Environment and Development, 1991. Our Common Future. Oxford University Press, Oxford.

Chapter 2

Living in a Bubble: A Mental Model of How the World Works

The bubble's not reality but it's inside your mind
Making you forget where you're from and what's behind...
The bubble is a very tricky thing all full of hype and it is not easy
To try to see the way things are they'll always be.

—Eiffel 65[1]

...most people in our modern society, and especially our large social institutions, subscribe to the concepts of an outdated worldview, a perception of reality inadequate for dealing with our overpopulated, globally interconnected world.

—Fritjof Capra and Pier Luigi Luisi (Capra and Luisi, 2014)

David Vetter lived most of his 12 short years inside a sterilized plastic bubble (see David in Fig. 2.1). Suffering from a severe immune deficiency disease, this handsome young boy could only feel his parents' touch through special gloves attached to the bubble's walls. Although he was schooled, watched television, and played inside his flexible confines, he could never experience or even breathe in the outside world. His perceptions of life were as small as his tiny space.

Though to a lesser degree, we all live in a bubble – a bubble created by our minds. From inside this bubble, we view the world narrowly and most often with respect to our own self-interest. But, our bubble is not plastic. It is built from our experiences, perceptions, and beliefs about how the world works. It dictates our choices and behaviors. This bubble is our mental model.[2]

---

1. Lyrics with permission of BlissCo from *Living in a Bubble* by Eiffel 65, Italian electronic dance group, written by Capuano, Lobina, Winston, Randone, and Ponte, © 1999 Bliss Corporation S.R.L.
2. In 1943, psychologist Kenneth Craik (1967) introduced the concept of "small-scale models of reality" that individuals use to anticipate events.

K.L. Higgins: Economic Growth and Sustainability. http://dx.doi.org/10.1016/B978-0-12-802204-7.00002-5

FIGURE 2.1 David Vetter, boy in the bubble. *Source: Photo courtesy of Baylor College of Medicine for use in this book, with approval of David Vetter's family.*

2.1 DEFINING OUR PREDOMINANT MENTAL MODEL

Mental models are fundamental to systems thinking. By holding an incomplete or inaccurate mental model, we may not appreciate the effects of our actions and may act in ways that cause damage. Although it is true, as systems expert Jay Forrester says, that "mental models are fuzzy, incomplete, and imprecisely stated," (Forrester, 1971) still, it is imperative to get at least a sense of what beliefs and perceptions are driving our worldly system. We want to understand how these fuzzy but powerful mental models undermine the goals of sustainability without our full awareness so that we can identify how to get there.

Therefore, with the knowledge that individual traits, cultures, experiences, and backgrounds create as many mental models of the world as there are people, we will take a bold leap and describe a mental model whose predominant beliefs influence the major forces that affect sustainability. We will hereafter refer to this model as "our mental model."

The model by no means characterizes the views of every individual; however, it does represent an aggregate of basic perceptions that dictate our behavior. As you read the book, keep this caveat in mind. From this model, we then piece together a more realistic picture of interactions and relationships among today's trends and events, and lay the foundation for deeper insight into sustainability.

2.1.1   Constraints of Individual Mental Models

Before we describe our mental model, let us first consider one constraining aspect of human nature that is common to us all. We humans seek the security of understanding our world and knowing what will happen to us. We cling to beliefs that bring order and establish patterns in the midst of uncertainty and ambiguity (Gilovich, 1991). Two characteristics of this pursuit of certainty dictate how we perceive the links among environment, economy, and society – the three components of sustainability.

First, past experiences bias our views of the future. Psychologists find that humans are predisposed to project current situations into the future.[3] We expect that our lives tomorrow will be just as they are today – or better. For sustainability, this mindset means that we believe we will awaken each morning to a world in which many of us have enough water to drink and food to eat, a world in which many of us depend on cell phones, automobiles, and electricity.

Second, mental limitations and human struggles constrict our field of concern to our own self-interest. We are endowed by nature with a survival instinct, one that necessarily focuses attention inward. We attend to "the things that are important to us," particularly those that arouse emotion, as neuroscientist Joseph LeDoux finds in his research on the human brain (LeDoux, 2002). This interest in self, often to the exclusion of others, relates directly to sustainability. Thomas Hobbes in the seventeenth century and B.F. Skinner in the twentieth century described this relationship well. They argued that to ensure survival of the fittest, human beings are by nature shortsighted and egoistic, thus any behavior that abuses the environment or takes it for granted is "simply an enactment of human nature."[4] In other words, as long as "I get what I need – enough clean air or water or money – I can ignore that my actions prevent others from getting what they need."

Counterarguments propose that such genetically based egoism is offset by our innate tendency to live in groups, which requires altruism and social learning. Nevertheless, the human predisposition toward individual or close-kin survival still governs behavior.[5] When we strive to survive, or to sustain our present lifestyles, or to find meaning, our brains gravitate toward short-term concerns and self-interest – to what affects us immediately and directly.

---

3. See Deci (1980). Economists use this psychological predisposition to understand human expectations about the market and the economy. See Shiller (2005).

4. Gardner and Stern (1996) refer to Hobbes and Skinner in their discussion of human dispositions as barriers to solving global environmental problems. See Thomas Hobbes' discussion of human nature and social contract in *Leviathan, or the Matter, Forme and Power of a Commonwealth Ecclesiasticall,* written in 1651 and Skinner (1971).

5. Gardner and Stern (1996); see also Wilson (1979).

2.1.2 Primary Beliefs in Our Mental Model

Our mental model emerges from this skeleton of short-term orientation and reliance on past experience. It is colored by a sense that Earth belongs to us rather than that we are part of the Earth. Three beliefs form the model's core: (1) economic growth fuels human thriving and will continue indefinitely; (2) technology advances will provide enough energy capacity to keep up with economic growth and to maintain our energy-dependent lifestyles; (3) population growth and pollution are beyond our concern because they do not directly or immediately affect our lives.

Although we could add a multitude of other beliefs, these three are fundamental. Together they explain why the global economy is still growing even though the effects of this growth profoundly damage the environment. They clarify why we consume energy as though carbon-based energy resources are infinite. They also explain (in part) why our global population continues to increase even though many nations cannot support their current inhabitants. With these three tacit beliefs, our mental model encases us in a small bubble where apprehension about the future is faceless and kept at arm's length.

In essence, our mental model asserts that economic growth, our seemingly stable environment, and our resources will go on forever. It can be likened to the "technocentric" perspective described by the United Kingdom's environmental specialists Michael Carley and Ian Christie. This view, which they assert is incompatible with sustainable development, "places faith in the capacity of technology to harness nature and substitute man-made capital for natural resources" Carley and Christie (2000).

2.2 ECONOMIC GROWTH AND HUMAN THRIVING

Of the sustainability triad, economy stands front and center; it has hijacked our attention and tipped the sustainability scale. We depend on economic growth for employment and sustenance. We rely on it to bring happiness by satiating our materialistic desires. Our symbiotic relationship with economic growth pervades every niche of human society; it is equally present in individuals and in collectives.

2.2.1 Individuals and Materialism

As individuals, many of us fret about keeping our jobs, about making payments on our credit cards, or about the price of milk. We want to buy new clothes, new cars, and new cell phones. Since the Greatest Happiness Principle in the late 1700s,[6] individual happiness has been equated to money and possessions; personal wealth symbolizes significance, achievement, security, and status. The perceived linkage

---

6. Utilitarians Jeremy Bentham and John Stuart Mill taught that morally correct actions are those that bring the greatest happiness to the greatest number; in its way, the principle justified capitalism to a frugal and ascetic society.

between money and happiness has strengthened in the past several decades. Nations (particularly those with high standards of living) have experienced "a shift from values emphasizing productiveness to values that promote consumption."[7]

Psychologist Philip Cushman finds that in the United States, for example, individuals now believe that material consumption brings "personal salvation, health, wealth, and popularity." Globally, excessive consumption exemplifies our human attempts to fill what Cushman calls a "pervasive sense of personal emptiness."[8] Whether we are rich or poor, whether we search for significance or survival, consumption is our Nirvana and the economy is our conduit.

To appreciate the magnitude and concentration of this consumption proclivity, consider that in 2006, 16% of the world's population accounted for 78% of its consumption and used 50% more nonrenewable resources than 30 years ago.[9] Furthermore, in the future, this materialistic penchant will not be a unique feature of advanced economies. It is oozing into the world's population like the *Blob*, the science fiction mass of goo that flowed down sidewalks and underneath doors to engulf unsuspecting residents.[10] As emerging economies are exposed to lifestyles of advanced nations through various media,[11] their populations aspire to join this buying behavior with hopes of raising their standards of living and their significance in the world. Imagine what will happen when the remaining 84% of world jumps on the materialism bandwagon.

2.2.2   Collectives and Economic Success

Individuals are not alone in their pursuit of economic success. Business organizations worry about revenue, expansion, and market share. They agonize about return-on-investment and they reward short-term performance. International stock markets demand profit and growth to attract shareholders.

Nations, too, concern themselves with the economy. They care about gross domestic product (GDP), inflation, unemployment, and balance of trade. Because GDP is a measure of global power, nations crave economic growth to elevate their stature, which in turn boosts their economies (AtKisson, 2012). Nations also rely on economic growth to support their populations; economic health is intricately linked to individual standards of living. As we discuss in later chapters, some nations are unwilling to invest in environmental cleanup when doing so reduces their GDPs, decreases available jobs, and diminishes the welfare of their citizens.

Individuals, organizations, and nations, all want the same thing: MORE.

---

7. For the United States, see Horowitz (1985).
8. Cushman (1995); Cushman calls this condition "the empty self."
9. Mattar (2012). Sixty-five high-income countries comprise the 16% of the world's population who account for 78% of the world's consumption.
10. *The Blob* is a 1958 sci-fi thriller written by Kay Linaker and Theodore Simonson and distributed by Paramount Pictures. (See <http://en.wikipedia.org/wiki/The_Blob>).
11. See Barnard (2012). Global advertising expenditures are projected to increase over 5% in 2014 above the estimated $525 billion spent in 2013.

2.2.3   World Economic Trends

The belief that economic growth is essential to human thriving has been forti-fied through experience. Other than a few dips, world GDP has climbed steadily since 1700.[12] After 2000, GDP growth accelerated, bolstered by an expanding population and more middle class consumers in a dozen emerging nations.[13] Most nations now boast higher GDPs than ever before. Global GDP growth between 2000 and 2012 was two and a half times more than that seen in the 20 years between 1980 and 2000.[14] In developing nations, China is the standout example of accelerated growth; since 1980, its GDP has more than *tripled* every decade.[15] Consumer spending has also increased. For example, from 2006 to 2011 in the midst of global recession, aggregate household spending rose 5% for the 34 countries in the Organisation for Economic Co-operation and Devel-opment (OECD).[16] Such escalation strengthens the belief that economic growth is not only possible and desirable, but that it is a law of nature just like gravity or survival of the fittest.

Of course, economic success in one country fuels economic growth in other countries. In some measure, all nations depend on it. Even the small Himalayan nation of Bhutan that coined the term "gross national happiness" has nudged its attention from the happiness of its citizens to economically driven goals that decrease unemployment and reduce national debt (Harris, 2013). Indeed, we are a world that is not only dependent upon but also *addicted* to growth to at-tain personal well-being and to meet the survival needs and material desires of our mushrooming global population. On the basis of our history with economic growth, we naturally expect it to continue.

2.3   ENERGY AND TECHNOLOGY ADVANCES

The second belief in our mental model rests on the premise that while we are dependent on energy, our ability to generate energy always has and always will meet our growing needs. Two assumptions motivate this belief. The first as-sumption is that Earth's carbon-based caches of fossil fuels (oil, natural gas, and coal) and its uranium deposits are abundant and will be around a "very" long time. And second, even if this earthly store is finite, technology advances will make energy usage more efficient, keep costs low, and add enough renewable

---

12. Maddison (2001). GDP data in 1990 International $ (Table B-18). Over the past thousand years population increased by a factor of 22, while world GDP increased by a factor of 300.

13. Court and Narasimhan (2010). These economies include the BRIC nations: Brazil, Russia, India, and China. In 2010, this slice of society numbered nearly two billion people.

14. International Monetary Fund (2013). World GDP average annual growth from 1980 to 2000 was $1,570B (purchasing power parity), while average annual growth from 2000 to 2012 was $4,033B.

15. International Monetary Fund (2013). China's GDP was $250B in 1980 and grew to $10,040B by 2010.

16. OECD (2013).

energy so that we may smoothly wean ourselves from fossil fuels at some inde-terminate point in the far distant future.[17]

2.3.1   Availability of Energy

The foundation of the belief that energy will always be there comes from experi-ence and from our discomfort with reality. Since the advent of electrical power transmission and gasoline-driven automobiles in the late nineteenth century, and the introduction of airplanes, telephones, appliances, and computers in the twentieth century, energy has been an integral part of our daily lives. Energy-powered devices that were earlier considered luxuries are now essential. Black-outs demonstrate just how dependent we are. No energy means no heat, no cooling, no lights; it means that restaurants and gas stations close, traffic stops moving, and computers and cell phones are inoperable. Trying to live without energy cripples us.

Our total reliance on energy and our complaisance with its presence are sub-stantiated by the worldwide trend of a rise in energy consumption. For example, as developing nations industrialized, consumption of fossil fuels between 1965 and 2012 more than tripled; in China alone, consumption increased 20-fold dur-ing this period. Use of fossil fuels is not only growing steadily, it is *accelerat-ing*. Fossil fuels produce energy and are the raw material for products ranging from DVDs and football helmets to antifreeze and lipstick. Between 1990 and 2000, consumption rose 13%. Between 2000 and 2010, it jumped by 28%.[18] These patterns illustrate that we not only depend on fossil fuels but also take them for granted – they are embedded in our cultures. We rarely consider that there is any limit at all.

2.3.2   Energy-Related Technology

The second tenet in our belief about energy is grounded in our success with technical advances. Because technology has solved our most challenging issues and has expanded our horizons, we naturally assume that it will somehow pres-ent itself at just the right moment to ensure that we never run short of energy. Decades ago, behavioral psychologist B.F. Skinner recognized technology's profound influence: "In trying to solve the terrifying problems that face us in the world today, we naturally turn to the things we do best. We play from strength, and our strength is science and technology" (Skinner, 1971). "Of course!" we say to ourselves.

---

17. Some distinguish between renewable and alternative energy and put both in the "green" cat-egory. In their view, renewable energy includes solar, geothermal, and wind power; alternative energy includes biodiesel, composting, and other organic ways to generate energy. This book lumps the two together as "renewable" energy and distinguishes between hydroelectric energy and other renewable energy.
18. Energy statistics from BP (2014).

FIGURE 2.2 Solar energy complex; San Bernadino County, California. *Source: Part of the 354 MW Solar Energy Generating System in the Mojave Desert. Photo courtesy of US Bureau of Land Management, public release. Retrieved from <http://en.wikipedia.org/wiki/File:Solar_Plant_kl.jpg>.*

Each day we see technology applied to conservation and production of energy. We have no reason to think that it cannot solve energy shortages in the future. Examples of energy efficiency technologies are commonplace. They range from heating systems to carpets made from recycled plastic to hybrid cars, whose popularity has recently expanded.[19]

On the production side, the number of patents in extractive industries doubled between 2005 and 2010.[20] More companies are using hydraulic fracturing (fracking) to remove oil and gas from previously inaccessible areas and are inventing new materials and machines to drill ultradeep underwater oil and gas wells. Renewable solar energy sources and wind turbines now dot many of our landscapes.[21] Since the 1980s when the first commercial solar panel plants were installed in the United States, solar complexes have appeared in countries such as India, Spain, and Germany. The largest solar power installation in the world transforms the sun's rays into electricity about 50 miles from my home in the Mojave Desert (Fig. 2.2). Its vastness and eye-watering reflections remind

19. In the United States, for example, cumulative sales of hybrids hit the 3 million mark in October 2013, Cobb (2013).
20. Elatab (2012). Extractive industries remove fossil fuels from the ground.
21. Two types of solar energy technologies are most prevalent. Solar thermal collectors use mirrors to concentrate and convert the sun's heat into steam that generates electricity. Solar panels have interconnected photovoltaic cells that convert the sun's energy directly into electricity.

passers-by of the immense potential for clean energy. Researchers continue to increase the efficiency of solar energy and reduce its costs.[22]

Such technologies lull us into trusting that all is okay on the energy front. And because we see such promise in energy technology, we dismiss its drawbacks. We ignore that fracking pollutes groundwater, depletes fresh water, and affects air quality (Brown, 2007). We do not want to know that in Colorado alone, fracking consumes enough water annually to meet the needs of over 118,000 homes[23] or that drilling accidents (e.g., the 2010 Deepwater Horizon oil spill in the Gulf of Mexico) harm wildlife, human health, and the environment. Since these accidents are infrequent and out of sight in most everyday lives, we discount them. Furthermore, our reliance on secondary energy sources masks our concern for the primary energy sources from which they are derived. In our minds, primary sources are simply inputs for the power grid or the gas station; they do not warrant individual attention.[24] Our energy needs will somehow be met.

2.4   POPULATION GROWTH AND POLLUTION

The human survival instinct presses us to avoid harmful circumstances and seek out opportunities to thrive. We are ever watchful and instinctively flinch when danger is imminent. However, if a threat does not seem to be immediate and harmful, or if we do not appreciate that it can hurt us at a future time, we disregard it and expend our energies on more urgent concerns. Growth of population and pollution fit this category; we push their potential for damage into the background. For now, although we may know that they will cause problems some time, population and pollution rest outside the boundaries of our mental model.

2.4.1   Population and Our Mental Model

Whether we live in cities or small towns, we can tell from crowded living quarters, freeway traffic, overflowing subway cars, water rationing, or public media that the number of people on Earth is increasing. It is hard to deny: In 2012 world population reached seven billion – an increase of one billion in only 12 years.[25] Although there are sporadic attempts, preventing population growth raises moral, legal, and political questions. To avoid engaging in this social dilemma, most of us regard more people as just another irritant.

In most regions, population growth has not yet dealt devastating penalties. In these regions, more people means more consumers – and this is good for business! In regions that *are* feeling the pressure of too many people, philoso-

---

22. Luque (2011). For example, researchers are improving voltaic cell efficiencies from 20% to 50%.

23. Belanger (2012). Note that this water is not recovered.

24. Secondary energy sources such as gasoline and electricity are derived from primary energy sources including geothermal, uranium, oil, natural gas, coal, sun, wind, biomass, and water.

25. Researcher Angus Maddison (2001) found that population in 1820 was just above 1 billion.

phies are different. Here, decision-makers focus on growing their economies to provide more food, more jobs, and higher standards of living even at the expense of the environment, as evidenced by the escalating pollution from coal-burning industries in China.

2.4.2 Pollution and Our Mental Model

The spread and concentration of pollution draws the same kind of disregard as population growth. We may hear that burning fossil fuels pollutes; that nuclear power plants create radioactive waste; that industrial processes release toxic chemicals and thermal pollution into air, water, and soil; and that humans produce waste and garbage. We may recognize that polluted water spreads disease, poisons flora and fauna, and eliminates fresh water supplies that sustain life. However, we do not want to know that much of the world's 400 billion tons of agricultural, industrial, and human waste finds its way into our rivers, oceans, lakes, and streams, and ultimately into our bodies (Rifai, 2013).

Except for a few places, such as Bejing where air pollution is an urgent health hazard or in West Java where the Citarum River is literally covered with untreated waste, and except for times when temperature inversions cause smog to accumulate in New York City or Bakersfield, California, we ignore these lifestyle byproducts and go on about our business.

Furthermore, as noted earlier, because we believe that economic growth is the key to human well-being, we favor it over costly attempts to clean the environment. A case in point involves the Kyoto Protocol that was established in 1997 to reduce greenhouse gas worldwide. In 1998, the United States researched implications of this policy and considered adding a "carbon price" to the price of energy to clean up the environment, postulating that "as energy costs rise, all consumers will tend to use less." The study predicted that if this tax were imposed, by 2010 GDP would drop by 1–4.2% (EIA, 1998). Unwilling to accept economic setback, the United States, China, and other nations did not ratify the protocol. Surprisingly, however, the United States was first to meet the Kyoto 2012 emissions target even though success resulted from a stagnant economy rather than from proactive measures (Watts, 2013).

Of course, not everyone ignores pollution. Chapter 12 describes significant progress in reducing toxic emissions. Technologies such as coal scrubbers that remove particulates from industrial smoke stacks and catalytic converters that remove toxins from automobile exhaust have been around for years. These advances remind us that technology can help us solve *some* problems. However, if solutions become too expensive, cost jobs, or affect economies, economy will trump environment and justify our indifference toward pollution and population growth.

This chapter introduces three fundamental beliefs about economic growth, technology, population, and pollution that form the basis of a mental model which is constrained by human traits of short-term focus and self-interest.

We must reiterate that, although this mental model does not reflect beliefs of specific individuals, nations, or cultures, it does represent the predominant behavior-shaping forces that govern world trends relating to sustainability. The next chapter translates this mental model into the visual language of systems thinking.

REFERENCES

AtKisson, A., 2012. Life beyond growth: alternatives and complements to GDP-measured growth as a framing concept for social progress. Tokyo, Japan: 2012 Annual Survey Report of the Institute for Studies in Happiness, Economy, and Society – ISHES. Retrieved from <http://www.isisacademy.com/wp-content/uploads/LifeBeyondGrowth.pdf>.

Barnard, J., 2012. ZenithOptimedia releases September 2012 advertising expenditure forecasts. ZenithOptimedia. Retrieved from <http://www.zenithoptimedia.com/zenith/zenithoptimedia-releases-september-2012-advertising-expenditure-forecasts/>.

BP, 2014, June. BP statistical review of world energy 2014. Retrieved from <http://www.bp.com/en/global/corporate/about-bp/energy-economics/statistical-review-of-world-energy.html>.

Belanger, L. 2012. Fracking our future: Measuring water & community impacts from hydraulic fracturing. Western Resource Advocates, Boulder. Retrieved from <http://westernresourceadvocates.org/frackwater/fracking_our_future_july_2012.pdf>

Brown, V., 2007, Industry issues: putting the heat on gas Environ. Health Perspect. 115 (2), A76.

Capra, F., Luisi, P.L., 2014. The Systems View of Life: A Unifying Vision. Cambridge University Press, Cambridge.

Carley, M., Christie, I., 2000. Managing Sustainable Development, second ed. Earthscan, London.

Cobb, J., 2013. Americans buy their 3,000,000th hybrid. Hybrid Cars. Retrieved from <http://www.hybridcars.com/americans-buy-their-3000000th-hybrid/>.

Court, D., Narasimhan, L., 2010. Capturing the world's emerging middle class. McKinsey Quarterly. Retrieved from <http://www.mckinsey.com/insights/consumer_and_retail/capturing_the_worlds_emerging_middle_class>.

Craik, K., 1967. The Nature of Explanation. Cambridge University Press, Cambridge.

Cushman, P., 1995. Constructing the Self, Constructing America. Addison-Wesley Publishing Co, Reading, MA.

Deci, E.L., 1980. The Psychology of Self-determination. Lexington Books, Lexington, MA.

EIA, 1998. Impacts of the Kyoto Protocol on U.S. energy markets & economic activity. U.S. Energy Information Administration. Retrieved from <http://www.eia.gov/oiaf/kyoto/kyotorpt.html>.

Elatab, M., 2012. 5 trends in oil & gas technology, and why you should care. Venture Beat. Retrieved from <http://venturebeat.com/2012/03/28/5-trends-in-oil-gas-technology-and-why-you-should-care/>.

Forrester, J., 1971. Counterintuitive behavior of social systems. Technology Review. Alumni Association of the Massachusetts Institute of Technology, Cambridge. Updated March 1995. Retrieved from <http://clexchange.org/ftp/documents/system-dynamics/SD1993-01CounterintuitiveBe.pdf>.

Gardner, G., Stern, P., 1996. Environmental Problems and Human Behavior. Allyn and Bacon, Boston.

Gilovich, T., 1991. How We Know What Isn't So: The Fallibility of Human Reason in Everyday Life. The Free Press, New York.

Harris, G., 2013. Index of happiness? Bhutan's new leader prefers more concrete goals. New York Times. Retrieved from <http://www.nytimes.com/2013/10/05/world/asia/index-of-happiness-bhutans-new-leader-prefers-more-concrete-goals.html?_r=0>.

Horowitz, D., 1985. The Morality of Spending: Attitudes toward the Consumer Society in America 1875–1940. Ivan R. Dee. Chicago.

International Monetary Fund, 2013. World Economic Outlook Database. Retrieved from <http://www.imf.org/external/pubs/ft/weo/2013/02/weodata/weoselagr.aspx>.

LeDoux, J., 2002. Synaptic Self: How Our Brains Become Who We Are. Penguin Group, New York.

Luque, A., 2011. Will we exceed 50% efficiency in photovoltaics? J. Appl. Phys., vol. 110, 031301.

Maddison, A., 2001. The World Economy: A Millennial Perspective. Organization for Economic Cooperation and Development, Danvers, MA.

Mattar, H., 2012. Public policies on more-ustainable consumption. In: Moving Toward Sustainable Prosperity. The Worldwatch Institute, Washington, pp. 137–144.

OECD, 2013. Statistics from A to Z. Organisation for Economic Co-operation and Development. Retrieved from <www.oecd.org/statistics/>.

Rifai, B., 2013, February 4. Pollution Facts Blog. Water pollution facts. Retrieved from <http://www.thepollutionfacts.com/2013/02/water-pollution-facts.html>.

Shiller, R., 2005. Irrational Exuberance, second ed. Princeton University Press, Princeton.

Skinner, B.F., 1971. Beyond Freedom & Dignity. Hackett Publishing Company, Indianapolis.

Watts, A., 2013. USA meets Kyoto protocol goal – without ever embracing it. Watts Up With That? Retrieved from <http://wattsupwiththat.com/2013/04/05/usa-meets-kyoto-protocol-without-ever-embracing-it/>.

Wilson, E.O., 1979. *On Human Nature*, Harvard University Press, Massachusetts, and Herbert Simon (1990). A mechanism for social selection and successful altruism. Science 250 (4988), pp. 1665–1668.

Chapter 3

The Ant Who Lives Forever: A Systems Interpretation of Our Mental Model

Working with systems...constantly reminds me of how incomplete my mental models are, how complex the world is, and how much I don't know.

—Donella Meadows (Meadows, 2001)

A hundred million years ago, the first ancestor of the ant family appeared on Earth (Brady et al., 2006). Today, the prolific presence of these tiny creatures makes them seem invincible and suggests that we can learn something from their endurance. Like this hardy insect that has survived for millennia, the myopic beliefs that form our mental model persist even as the world around us whirls with change.

Yet, the ant has more in common with our mental model than persistence. We find a physical and behavioral similarity when we translate our mental model into the visual language of systems thinking. To picture this resemblance, take a peek at the system diagram in Fig. 3.2 and mentally superimpose it onto the picture of the carpenter ant in Fig. 3.1. Just as the ant consumes food and produces waste, oblivious to the rest of the world, our mental model motivates us to consume excessive energy and produce piles of pollutants without regard to the effects of our behaviors.

However, the ant differs from our mental model in a fundamental way; it has adapted to massive environmental change, while our beliefs inhibit our ability to adjust as successfully. This chapter illustrates that clinging to our simplistic mindset opposes sustainability. We must, therefore, reshape perceptions and behaviors to accommodate the rising complexity of our world and to cope with impending global change. Later in the book, Chapter 8 takes on this transformation challenge and converts the ant-like mental model into a more elaborate system diagram with multiple interactions and influences.

3.1 SYSTEMS DEPICTION OF OUR MENTAL MODEL

Like tree roots, three sustainability-related beliefs have embedded themselves in our mental model of the world, curling and twisting to avoid the rocks of reality. Although these predominant beliefs (see Chapter 2) may rest beneath our

K.L. Higgins: Economic Growth and Sustainability. http://dx.doi.org/10.1016/B978-0-12-802204-7.00003-7

FIGURE 3.1 Carpenter ant.[1]

threshold of consciousness, they shape individual behaviors that define global trends.

First, we are certain that economic growth is essential to our well-being and that it will always continue. Second, we believe our sources of energy are boundless since technology guarantees that our needs will always be met. And finally, we avoid thinking about population growth and pollution; they are someone else's worry. By disregarding the consequences of these beliefs, we live comfortably and complaisantly, reassured by the voice in our heads that says everything will be all right.

Now let us look more deeply at our mental model and examine how its three fundamental beliefs interact. We call upon the feedback loops and arrows of systems thinking to translate the relationships among these beliefs into the system diagram in Fig. 3.2. The three reinforcing loops in this ant-like diagram gain momentum from one another's influence. Like the interlocking gears in Chapter 1, when one loop spins, others also spin so that we cannot distinguish what triggered their movement. To interpret the diagram, recall that short-term focus and past experience underpin our mental model and define the system boundary (delineated with dashed lines). Recall also that an arrow with an "s" (same) indicates that the two elements it connects (cause and consequence) move up or down together and that an "o" (opposite) means that when cause goes up, consequence goes down and vice versa.

The following descriptions locate each of the three beliefs on the system diagram. For clarity, this discussion italicizes the names of elements and loops as they appear in the diagram.

3.1.1 Economic Growth and Personal Gratification

Two reinforcing loops, the *economy-reinforcing loop* and the *gratification-reinforcing loop*, illustrate the first belief that fuels our mental model. They represent the economic growth engine that gratifies our materialistic desires. The foundation of our mental model, the *economy-reinforcing loop* in the center of the diagram reflects the incredible economic growth that is occurring today. In

---

1. Carpenter ant drawn by California artist Linda Anfora; reproduced with permission of the Western Exterminator Company.

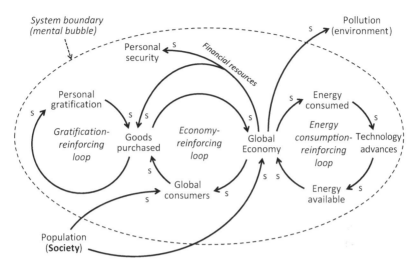

FIGURE 3.2   Mental model system diagram.

this loop, as *the global economy* grows, more people have the ability to buy more products, thus economic growth produces more *global consumers.* More *global consumers* means more *goods purchased* and more *goods purchased* increases the *global economy* – a true growth cycle. Finally, when the *global economy* grows, more people have *financial resources* and feel greater *personal security.*

In the *gratification-reinforcing loop* on the left, more *goods purchased* raises *personal gratification.* Growth in the *global economy* means that more *global consumers* purchase goods to increase their *personal gratification.* However, *personal gratification* is short-lived and must constantly be fed, that is, *goods purchased* will continue to rise as we satiate our material desires. This last connection completes the reinforcing loop. Psychologists call this loop the "hedonic treadmill" and systems thinkers call it a "vicious circle": when individuals pursue a materialistic path toward happiness, the more they have, the less fulfilled they feel, and the more they want.[2]

3.1.2   Abundant Energy and Technology Advances

The rightmost loop in Fig. 3.2 incorporates the second belief in the mental model: We act as though technology advances and abundant energy resources will forever grow our economy and improve our lifestyles. In this *energy-consumption-reinforcing loop*, growth of the *global economy* increases the amount of *energy consumed* for designing, manufacturing, and distributing goods, and providing services. When *energy consumed* rises, energy-related *technology advances*

---

2. The "hedonic treadmill" was first coined by Brickman and Campbell (1971); see also Ariely (2010).

must also increase to keep up with the need for innovative ways to generate energy. More *technology advances* raise *energy available* that induces growth of the *global economy*. In this loop, because we believe energy is ever present, neither the threat of an energy shortage nor the fear that we are consuming too much energy inhibits economic growth. For example, if we acknowledged that we are fast depleting our fossil fuels or if energy prices were higher, we would be more inclined to conserve. However, we are either ignorant of the effects of our behavior or ignore these effects since reducing consumption would also impair our lifestyles and depress the economy.

3.1.3 Population and Pollution Are External

The third tacit belief places *population* and *pollution* outside the system boundary and beyond our immediate concern. They are simply an input and an output of the growth machine. In the diagram, increased *population* contributes to economic growth in two ways. First, it adds to the pool of *global consumers* who foster economic growth. Second, it drives economic growth directly since there are more people to productively contribute and nations expand their economies to support accumulating numbers of people who need jobs and sustenance. In this depiction, a decline in *population* – which has not occurred since the Middle Ages – would depress the economy. Otherwise, *population* is irrelevant and unaffected by other elements.

Pollution is only a byproduct of the *global economy*. We know that when our economic engine is running, it will belch smoke and smog; we also know that as we become more affluent, we buy more stuff and produce more garbage. In the diagram, however, this pollution does not reenter the system as feedback that will create consequences.

Although some individuals and organizations have recognized the negative effects of population growth and pollution, and are attempting to diminish these effects, their scattered efforts are a drop in the ocean compared to the magnitude of the problem. So, whether population is escalating in Bangkok, Delhi, or Lagos, or pollution is amassing in the Citarum River in Indonesia, the Matana–Riacheulo River in Argentina, or the Mississippi River in the United States, the truth is that because most of us have not yet felt substantial damage, we do not worry much about population or pollution. We just "kick the can" into the future.

3.2 IMPLICATIONS OF OUR MENTAL MODEL

As Chapter 4 will show, statistics confirm that our mental model accurately reflects some aspects of current reality: Both the global economy and energy consumption have grown to unprecedented heights. Technology to increase renewable energy and improve energy efficiency is on the upswing, and for the most part allows us to bury our angst about fossil fuels. Escalating population and pollution are their own problems whose harmful effects are localized and whose solution comes only at the expense of economic growth. Finally, increased consumer spending and advertising blitzes tell us that once we have reached a level of subsistence, countless numbers of us find ourselves running on the hedonic treadmill in search of instant gratification. Chapter 5 expands this discussion of how we seek happiness.

3.2.1   Unbounded Growth

With this trio of mutually energizing reinforcing loops, we have envisioned a system that can grow without bounds: consumption of energy, economy, and personal gratification will rise forever. Thus we are inclined to make few changes, trusting that this glorious state can continue without exceeding the energy resources required to sustain it and without experiencing too much damage from pollutants.

However, systems thinking tells us that growth cannot go unchecked; it inevitably reaches its *limits-to-growth* as Chapter 1 discussed. At these limits, the behavior of a system will either placidly stop growing, or severely react and collapse,[3] as we will further discuss in Chapter 7. Pulitzer Prize recipient Jared Diamond clearly recognizes these two paths for pollution, noting that "the world's environmental problems *will* get resolved ... The only question is whether they will become resolved in pleasant ways of our own choice, or in unpleasant ways not of our choice" (Diamond, 2005).

3.2.2   Unrealistic Beliefs

Our mental model unrealistically ignores both the soft-landing and brutal-collapse options of how growth can end; in fact it fails to acknowledge any end to growth at all. Because we tend to consider short-term effects and suppress nagging concerns about more people and more waste on the planet, we neglect the long-term consequences of unrestrained growth. Furthermore, sustainability challenges are so encompassing that they seem beyond the influence of one person. Among these challenges is the social dilemma called "tragedy of the commons" in which individuals act in their own self-interest relative to a common resource (e.g., the atmosphere or nonrenewable fossil fuels) and ultimately compromise the long-term best interests of the common good.[4] In this situation, if any individual reduces his consumption, he will only be hurt while others profit from his pain (Chapter 14 expands on this phenomenon). Thus, human

---

3. The first condition is called "s-shaped growth" and the latter condition in which growth exceeds capacity and then drops abruptly is called "overshoot-and-collapse" (see Sterman [2000] and Higgins [2013]).
4. Hardin (1968) first coined this term using the analogy of common pastureland shared by individual farmers.

nature, experience, culture, and a sense of impotence combine to justify our actions and inaction.

It is not surprising that we search no further for the rest of the story; it is too frightening and too remote. We are comfortable living within the confines of our mental bubble. The system diagram of our ant-like mental model reflects a simplistic world in which worries are not as overwhelming as reality.

However, the world is more complicated. The next few chapters will burst the myopic bubble that protects us from concern and keeps us on the path of excess and growth. Using historic data and future projections, these chapters help us view economic growth differently, expand our perspective of population and pollution, and explore limits to growth. The ant's behavior will take on more human dimensions.

REFERENCES

Ariely, D., 2010. The Upside of Irrationality. HarperCollins, New York.

Brady, S., Schultz, T., Fisher, B., Ward, P., 2006. Evaluating alternative hypotheses for the early evolution and diversification of ants. PNAS 103 (48), 18172–18177.

Brickman, P., Campbell, D., 1971. Hedonic relativism and planning the good society. In: Appley, M.H. (Ed.), Adaptation-Level Theory. Academic Press, New York.

Diamond, J., 2005. Collapse: How Societies Choose to Fail or Succeed. Penguin, New York.

Hardin, G., 1968. The tragedy of the commons. Science 162, 1243–1248.

Higgins, K., 2013. Financial Whirlpools: A Systems Story of the Great Global Recession. Elsevier, Oxford.

Meadows, D.H., 2001. Dancing with systems. Whole Earth. Retrieved from <http://www.wholeearth.com/issue/2106/article/2/dancing.with.systems>.

Sterman, J., 2000. Business Dynamics: Systems Thinking and Modeling for a Complex World. Irwin McGraw-Hill, New York.

Chapter 4

Addicted to Growth: Economic Growth Promises Happiness and Well-Being

Why are so many economies dependent on consumption elsewhere? Their dependence stems from the path they chose toward rapid growth...

– Raghuram Rajan (Rajan, 2010)

But despite the best efforts of philosophers and economists over the centuries to attribute their operating assumptions to the same laws that govern nature, economic paradigms are just human constructs, not natural phenomena.

– Jeremy Rifkin (Rifkin, 2014)

It is easy to view the future as a repeat of the past – it is what we humans do well. We project what we are used to seeing and trust that the world will remain as it is. Every day our mental model interprets our surroundings and rationalizes current trends to fit within its confines. Every day we unknowingly struggle to hold on to these beliefs about how the world works – to do otherwise would shake our foundations. Just like touching the elusive reflections of sunlight shining through a crystal, we grasp at the glimmers of hope built from the belief that economic growth is a law of nature and will forever create social well-being and happiness.

As we discovered in the previous two chapters, the central tenet of the mental model we protect so closely rests on continuous economic growth that must be fueled by infinite supplies of energy, endless desire for material goods, and countless consumers. National economies have grown substantially since the Industrial Revolution when innovative technology, new materials, and inventive processes increased productivity worldwide. As a result, humankind has reaped the benefits of greater food production, fancy work-saving devices, stuff that we did not know we needed, and ultimately a higher quality of life. In most regions, we like our place and time in history!

However, limits are missing in this flywheel of ever-rising production and consumption; our mental model foolishly disregards, as cultural evolution expert Andrew Schmookler says "...the value of what the earth provides

K.L. Higgins: Economic Growth and Sustainability. http://dx.doi.org/10.1016/B978-0-12-802204-7.00004-9

31

us [and] subsidizes the recklessness of the present at the cost of the future" (Schmookler, 1993). Like shoppers who place apples in their grocery bags with no thought of the orchards, farmers, water, fertilizer, machines, or trucks involved in getting them to the store, we have become accustomed to the fruits of economic growth without considering how they are produced. This complaisant behavior is not merely habit. It is addiction.

4.1 ADDICTED TO ECONOMIC GROWTH

Habits take a few weeks to break. The more they are ingrained in our neural circuitry, the longer it takes to erase them from our repertoire. Addictions are different. While habits involve choice – we choose habitual behaviors and can choose to stop if we work at it – addiction has both psychological and physical components. Feeding an addiction often requires irrational or self-damaging behavior. Addicts who want to end their dependence – whether on caffeine, or alcohol, or drugs, or food – find themselves hard-pressed. And herein is the key to understanding how extensively economic growth permeates our mental model. As though bound by the ball and chain in Fig. 4.1, we are as tied to economic growth as we would be to any addictive substance.

Why is economic growth so enticing? First, it is personal. It connects to our human nature and our physical makeup. Chapter 3 described the need to consume as being on a *hedonic treadmill* – the never-ending, never-satiated desire for happiness through material goods. When individuals believe that money and things will make them happy, they cannot step off this treadmill; the more they have, the less fulfilled they feel and the more they want. As individuals who are caught on it, we buy products. As organizations, we want more consumers to buy our products. As nations, we rely on the economic growth that is stimulated by all this buying to provide jobs and embellish our influence.

Nicotine, caffeine, drugs, chocolate, food, alcohol, things ... and economic growth

FIGURE 4.1 Economic growth and addiction.

The second reason we love economic growth is cultural. The embedded values in a culture are formed in part by what an organization or a nation measures and rewards. Most nations measure their well-being, that is, their power, wealth, and success, in terms of the gross domestic product (GDP) of their economies. Business organizations measure success in terms of profit, return on investment, and growth. It is true, however, that a few nations have unique cultures because their metrics and values place their economies at lower priority than their people. For example, tiny Bhutan in the Himalayas measures happiness as material *and* spiritual development. The Pacific archipelago Vanuatu, because of its emphasis on human and environmental well-being, ranks first on the *Happy Planet Index.*[1] In spite of these few exceptions, with a push from the many cultures that revere economic success, individuals in most modern-day countries equate happiness with wealth and material goods. They seek money and things rather than satisfaction derived from other sources.

A final reason economic growth is so important to us is societal. A growing economy is a time-tested way to maintain or increase standards of living, to improve our status in the world, and to cure social ills. It takes more jobs, more businesses, more infrastructure, and more money to support more people in the same-sized world. Together, these factors have fueled unparalleled expansion of the global economy.

4.2   ECONOMIC GROWTH TRENDS

The ant-like diagram of our mental model in Chapter 3 feeds on consumers whose search for personal gratification stimulates economic growth; the more goods they purchase, the more energy and technology it takes to create these goods. Economic growth generates more economic growth. After all this processing, the system produces waste and begins all over again – simple concept, simplistic mental model. Historic and projected trends corroborate these straightforward relationships: consumer spending is increasing, the global economy is growing, and world population increases by one person every 13 seconds.[2]

4.2.1   Population, GDP, and Consumer Spending Compared

Fig. 4.2 compares growth trends for population, GDP, and consumer spending. The chart shows that about 1.7 billion people will add to world population between 2015 and 2040. That is about 1¼ times the number of people in all of China today. Because more GDP is required to maintain the standard of living for more people, the world economy (GDP) rises as population grows.

However, we can see that the *rate* of GDP growth has far outpaced the *rate* of population growth. Between 1960 and 2040, population triples while

---

1. Boyle and Simms (2009). Bhutan uses gross national happiness instead of gross national product. The Happy Planet Index measures human well-being and low environmental impact.

2. Retrieved from <http://www.census.gov/popclock/>.

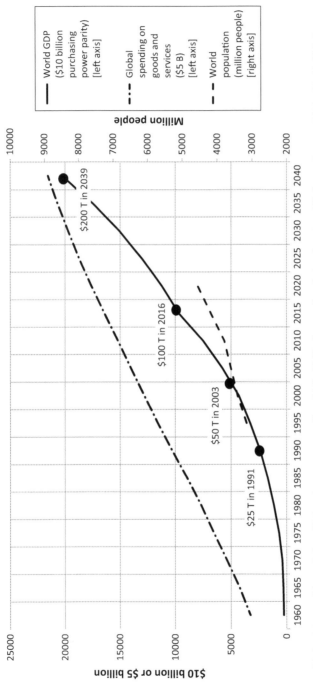

FIGURE 4.2 World population, GDP, and consumer spending (1960–2040). *Source: Population data from US Census Bureau (2013). World GDP from 1980 to 2015: International Monetary Fund (2013). For 1970 and 1975: International Monetary Fund (1999). GDP growth projected at 3% per year from PricewaterhouseCoopers (2013). World GDP for 1960 and 1965 were extrapolated from World Bank chart. Retrieved from <http://profitfromfolly.files.word-press.com/2012/07/world-gdp-1960-2010.png>. Global spending on goods and services: Kearney (2012). GDP and global spending data are expressed as purchasing power parity. Note the $10 billion scale for GDP and the $5 billion scale for global spending.*

GDP increases by a factor of 80. Therefore, population growth does not fully explain the rise in GDP. In the figure, GDP growth has accelerated since the 1960s, about the same time that digital technology surfaced. After electronic networks matured in the 1970s, personal computers hit the scene in the 1980s, the World Wide Web emerged in the 1990s, and search engines such as Google and Yahoo! appeared in the 2000s. Doing business has never been easier. People around the world not only conduct business remotely but also order products from other countries with no fuss. With today's smart phones and computer tablets we can spend our money instantly from anywhere whenever the urge strikes. Furthermore, access to an enormous amount of information via the Internet has sparked innovation and technical advances that also feed the economy. In addition to an increased number of consumers, all these technological factors contributed to the unprecedented upsurge in the global economy.

As our networked world grows new connecting fibers of communication and transportation, researchers suggest that we are in the midst of a third "super-cycle" (Standard Chartered Bank, 2010). One implication of this super-cycle is that GDP will rise about 3% a year between 2015 and 2040, bringing our spending power with it (PricewaterhouseCoopers, 2013). Of course, as world GDP climbs, so does consumer spending (and vice versa) particularly in middle class households. Between 2010 and 2020, global consumption of goods and services should increase 43% from $28 trillion to about $40 trillion as GDP jumps 50% (see Fig. 4.2). Half this increase will be in emerging economies such as China, India, and Brazil.

4.2.2   Implications for Standards of Living

From these aggregated statistics, we might be tempted to conclude that because population is growing more slowly than GDP, everyone will have a bigger share of the economic pie, that is, all standards of living will increase. While this conclusion may be true for some countries, it does not apply to others.   '

One reason for this disparity is that economic growth among nations will be more uneven than ever. Huge emerging economies will displace established economies. China, for example, is poised to overtake the United States as the world's largest economy by 2017 (PricewaterhouseCoopers, 2013). Fig. 4.3 shows that between 2011 and 2030, China's GDP will nearly triple. India's GDP will more than triple. The GDPs of Brazil and Mexico will also double, while advanced economies will experience only moderate growth.

For countries such as China, whose 2012 GDP growth (7.8%) far outpaced its population growth (0.5%), these statistics mean that the standard of living will improve for many of its people. However, countries such as South Sudan (whose 2012 GDP dropped by half while its population increased over 4%) will face especially difficult times; if these trends continue, the Sudanese will suffer a marked decline in standard of living.

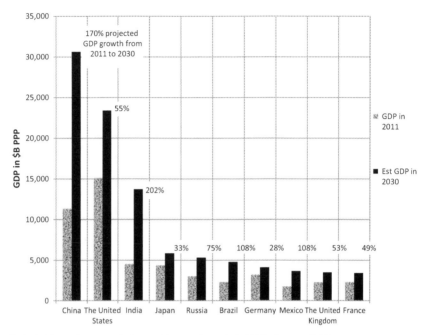

FIGURE 4.3 GDP growth for top 10 countries (2011–2030). *Source: GDP in purchasing power parity from PricewaterhouseCoopers (2013).*

Other factors are at play as well. Even Africa's largest growing economy, Nigeria, will be challenged. With its nearly 7% GDP growth but less than 3% population increase in 2012, one would expect that conditions for Nigeria's people were improving. However, its staggeringly high 30% unemployment rate in 2013 is predicted to increase 2% in 2014.[3] Countries that experience high unemployment like Nigeria will also suffer enormous income inequality; fewer people will have the bulk of the wealth. We will discuss income inequality in Chapter 5.

4.2.3 Short-Term Benefits

National economies are linked to the expanding global economy. Short-term benefits of economic growth seem obvious: The more businesses and nations that profit, the more individuals have jobs and resources and the higher their standards of living. More people with jobs and resources consume more goods and services. More goods and services consumed feed the economic machine and the cycle continues.

---

3. Chima (2014). Nigeria's GDP growth was 6.7% and its population increase was 2.8% in 2012. Its rate of unemployment was 23.9% in 2011 and its urban unemployment was an estimated 29.5% in 2013.

Yet even with its advantages, economic growth has also become a murky mantle, adorned with excessive debt, unhappiness, and devastation. For many who know the hidden costs of this mantle, concern has developed into anxiety about how to avoid a bleak present and an even bleaker future. Diamond (2005) nicely summarizes the conundrum of economic growth: "What makes money for a business, at least in the short run, may be harmful for society as a whole." Still, we might ask: How can growth possibly be harmful? To answer this question, let us consider how the byproducts of economic growth damage our world.

4.3   REPERCUSSIONS OF ECONOMIC GROWTH

Recognizing the repercussions of economic growth is relatively contemporary. National economies have flourished in the past three centuries since Scottish philosopher Adam Smith recognized that laws of supply and demand create a self-governing market (Smith, 1937). In the late twentieth century, environmentalists and others began to challenge this laissez-faire view of the economy and question our dependence on economic growth.

In addition to improving quality of life and allowing more people to pursue happiness by buying material goods, economic growth consumes precious natural resources and emits pollutants that threaten the delicate climate on which life relies. Thus, there are limits to economic growth and to our consumption-centric lifestyles. But what do these limits mean?

If economies expand too much, they will deplete the energy supplies before alternatives are available. If economies expand too much, the pollution produced by burning fossil fuels, growing more food, and discarding our garbage will undermine the very foundation of economic growth: society. Chapter 6 describes these threats to growth in more detail. It assesses the availability of energy and explains how pollution chokes air, water, and land, and creates harsh weather patterns that affect food, water, and well-being; finding alternative energy sources and repairing this damage costs money.

Consumptive behaviors dictated by our mental model have become a way of life. Our addiction makes us forget there are limits to the resources that facilitate the way we live. These behaviors are unsustainable. As noted ecological economist Tim Jackson has said "The vision of social progress that drives us – based on the continual expansion of material wants – is fundamentally untenable. ...In pursuit of the good life today, we are systematically eroding the basis for well-being tomorrow" (Jackson, 2011).

Yet, it will be difficult for nations to adjust to economic decline and for individuals to make fewer purchases and accept lower standards of living. We face a real dilemma. Just ask yourself: How hard would it be to stop buying things beyond what I need to survive? How tough would it be to stop driving my car or watering my lawn? What if my electricity were shut off for days or weeks at a time? Toward the end of the book we will explore a middle ground between having everything now but nothing in the future, and saving everything for the future but having nothing now.

4.4 A PROMISE BROKEN: CREATING A NEW PERSPECTIVE

Before recommending a solution to the dilemma between present and future, we must first understand what is happening in our world and how economic growth affects sustainability. To accomplish this feat, we need a fresh perspective that recalibrates our mental model. This perspective must recognize that over the long term, economic growth cannot meet its promise to make us happy. This new view must incorporate more potent ways to realize human happiness and well-being than instant personal gratification. It must revise the boundaries of our model and integrate the effects of increasing population and pollution. Finally, it must acknowledge long-term outcomes and incorporate factors that limit growth.

This chapter described why we are addicted to economic growth: it promises happiness through endless consumption and pledges well-being through increased quality of life. If the rate of economic growth could continue forever, and if all people benefited equally, this addiction would be a pleasant one. However, ... economic growth cannot continue. In the long run it will reach its limits and its many repercussions will wreak havoc on society. Because these consequences do not appear in our mental model, we tend to ignore them. Thus, we must exchange our current beliefs for new ones.

With this goal in mind, Chapters 5–7 identify alternative sources of happiness beyond personal gratification, introduce the harmful effects of economic and population growth, and incorporate growth-limiting factors. Chapter 8 wraps these insights into a neat bundle and creates an integrated system diagram of how the world works relative to sustainability's major influencers. This holistic view opens our eyes and allows us to identify lasting solutions.

REFERENCES

Boyle, D., Simms, A., 2009. The New Economics: A Bigger Picture. Earthscan, London.

Chima, O., 2014, January 7. Nigeria's unemployment rate may rise by 2%. This Day Live. Retrieved from <http://www.thisdaylive.com/articles/nigeria-s-unemployment-rate-may-rise-by-2-/168227/>.

Diamond, J., 2005. Collapse: How Societies Choose to Fail or Succeed. Penguin, New York.

International Monetary Fund, 1999, September. World economic database. Retrieved from <http://www.imf.org/external/pubs/ft/weo/1999/02/data/>.

International Monetary Fund, 2013, October. World economic database. Retrieved from <http://www.imf.org/external/pubs/ft/weo/2013/02/weodata/weoselagr.aspx>.

Jackson, T., 2011. Prosperity without Growth: Economics for a Finite Planet. Earthscan, London.

Kearney A.T., May 2012. Consumer wealth and spending: the $12 trillion opportunity. A.T. Kearney Global Consumer Institute: Ideas and Insights. Retrieved from <http://www.atkearney.com/paper/-/asset_publisher/dVxv4Hz2h8bS/content/consumer-wealth-and-spending-the-12-trillion-opportunity/10192>.

PricewaterhouseCoopers, 2013, January. World in 2050. The BRICs and beyond: prospects, challenges and opportunities. Retrieved from <http://www.pwc.com/en_GX/gx/world-2050/assets/pwc-world-in-2050-report-january-2013.pdf>.

Rajan, R., 2010. Fault Lines: How Hidden Fractures Still Threaten the World Economy. Princeton University Press, Princeton.

Rifkin, J., 2014. The Zero Marginal Cost Society: The Internet of Things, the Collaborative Commons, and the Eclipse of Capitalism. Palgrave Macmillan, New York.

Schmookler, A., 1993. The Illusion of Choice: How the Market Economy Shapes Our Destiny. State University of New York Press, New York.

Smith, A., 1937. An Inquiry Into the Nature and Causes of the Wealth of Nations. Random House, New York.

Standard Chartered Bank, 2010, November 15. Global Research: The Super-Cycle Report. Retrieved from <http://www.standardchartered.co.id/_documents/press-releases/en/The%20Super-cycle%20Report-12112010-final.pdf>.

US Census Bureau, 2013. World population: total midyear population for the world: 1950–2050. Retrieved from <http://www.census.gov/population/international/data/worldpop/table_population.php>.

Chapter 5

Two Faces of Happiness: Instant Gratification versus Sustainable Well-being

Our happiness depends at least as much on our environment and our culture as on our genes...

– Stefan Klein (Klein, 2002)

This is the true joy in life, the being used for a purpose recognized by yourself as a mighty one; ... the being a force of Nature instead of a feverish selfish little clod of ailments and grievances complaining that the world will not devote itself to making you happy.

–George Bernard Shaw (Shaw, 2000)

How would you complete this sentence? I am happy when I _____.

Was your answer "buy a new pair of shoes" or "get a new iPad?" Or was it "hold a newborn," "hike in the mountains," "spend time with family," or "learn something new?" Perhaps your response had two parts: one that acknowledges the immediate buzz from acquiring something and another that revels in the deep satisfaction of belonging, sharing, or enjoying – feelings which can neither be purchased nor consumed. Like Janus, the Roman god of transitions, happiness has two faces (see Fig. 5.1). While Janus' faces look outward in opposite directions to the future and to the past, one face of happiness seeks instant gratification and the other searches for sustainable well-being.

5.1 BLENDING EASTERN AND WESTERN IDEALS

The system diagram of our mental model from Chapter 3 includes only the first face of happiness: *personal gratification*. Its short-term perspective focuses narrowly on the individual. This face is self-serving and fleeting; it relies on some form of the hedonic treadmill. By tying security to money, and happiness to consumption of goods, it creates a *gratification-reinforcing loop*. This trait of human nature, the proclivity for consumption and short-term gratification, is reinforced by Western philosophies that foster individualism and self-centered pursuits. It grows stronger with increasing affluence.

K.L. Higgins: Economic Growth and Sustainability. http://dx.doi.org/10.1016/B978-0-12-802204-7.00005-0

FIGURE 5.1   Janus. *Source: Janus: Roman coin. (Photograph).* Encyclopædia Britannica Online. *Retrieved from <http://www.britannica.com/EBchecked/media/6135/The-god-Janus-beardless-Roman-coin-in-the-Bibliotheque-Nationale>. Credit: Larousse.*

The second, sustainable, face of happiness leads to a more fulfilled life. Its roots reflect Eastern philosophies that cherish relationships and look for meaning in life within the context of a greater whole. To incorporate this more holistic and long-term notion, we must expand the dominant world view.

5.2   SUSTAINABLE HAPPINESS AND WELL-BEING

This chapter describes two of the many factors that influence the second face of happiness: (1) income equality which gives us significance; and (2) relationships, health, and meaning which fulfill our deepest yearnings. Including this face in our systems perspective will help us resolve the dilemma between economic growth with its emphasis on the present, and sustainability with its eyes on the future.

5.2.1   Income Inequality

Numerous studies contradict our mental model's version of happiness; they debunk the link between happiness and money (and the things that money can buy). In hundreds of international surveys conducted since World War II and in research on happiness and well-being, all agree that "money brings satisfaction, but the effect is minimal" (Klein, 2002). Having money produces happiness only when one's income falls below a subsistence level; there is no connection when one's income rises above that level.[1] Richard Heinberg from the Post Carbon Institute says it well: "The economic treadmill is continually speeding up ... and we have to push ourselves ever harder to keep up, yet we're no happier as a result" (Heinberg, 2011). The treadmill is definitely not the best path to well-being.

---

1. See Edwards (2010). Conclusion is based on several studies and on international satisfaction level data from the World Value Survey, Frey and Stutzer (2002). Note that in 1995 this threshold was $10,000. Hernández-Murillo and Martinek (2010) found that mean happiness in the United States was relatively flat, while per capita income (in 2005$) doubled.

But this is not the end of the argument for money and goods. People care about how they compare with others and feel miserable if they do not measure up to their peers. Reasons for such behavior range from *equity theory* that is based on fairness,[2] to *moral* theories of justice and equality that assess how individuals are treated relative to one another (Frankena, 1973).

Whether we are dealing with civil rights, social status, clothing, or income, humans want to be treated fairly and compare favorably with others. We can easily identify this tendency in children; if you spend time with young ones, at some point you will see hurt feelings or anger when they are treated differently.

With regard to income, research tells us that at the individual level "absolute income level has little effect on happiness, [but that]... relative income matters: The lower the income of the group one compares oneself with, the more satisfied people are."[3] Schmookler suggests why this relationship is true: "Wealth would appear to be an arena of interpersonal competition for status (and also for power)" (Schmookler, 1993). We need to feel significant.

Rather than being the sweet elixir of happiness, money and income have a bitter side. Global trends reflect a growing inequality of income both *within* and *between* nations – a disparity that touches the heart of human discontent. In emerging economies that expect unprecedented growth, individuals' incomes will still be well below those in advanced economies. For instance, between 2010 and 2050, the average Indian's income will increase from one-thirtieth to about one-third the income of an average Brit; the average Chinese worker's annual income will grow from $2,200 to $31,000, or about half of that in the United Kingdom and less than two-fifths of that in the United States.[4] This inequality among nations will still create the longing to compare favorably with others, and will ultimately diminish happiness.

Income inequality *within* nations (measured by the Gini index) echoes similar disparity.[5] A higher Gini index means a greater inequality. This index has steadily increased since 1982 in all nations except China and India (Conference Board of Canada, 2013) where it has risen only in the past few years (Gore, 2013). Fig. 5.2 shows that African and South American countries have the greatest income inequality, that is, a few people receive most of the wealth. As we suspected in Chapter 4, income inequality explains why unemployment is so high in Nigeria even though its GDP is growing faster than its population. In Germany and Sweden

---

2. Mowday (1991). Equity theory says that people are satisfied when they believe that the value they receive from their labor equals the effort they expend, particularly relative to others in similar circumstances.

3. Frey and Stutzer (2002); see also Offer (2007) and Klein (2002).

4. Smith (2010); 2050 average income in the United Kingdom estimated at $59,000 and in the United States at $83,000. Comprehensive statistics are available from the Luxembourg Income Study database at <http://www.lisdatacenter.org/>.

5. The Gini index measures the income distribution of a nation's residents. A value of 0 means everyone has the same income; 100 means one person has all the income.

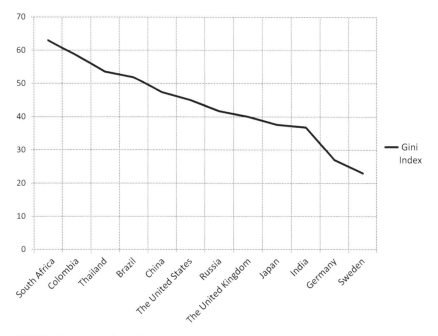

FIGURE 5.2 Income inequality (Gini Index) for selected countries. *Source: CIA (2014); most recent data available was used.*

more people share the wealth. Surprisingly, income inequality in China is only slightly higher than income inequality in the United States.[6]

Nobel Prize recipient Joseph Stiglitz recently highlighted the ascent of income inequality in the United States: in the past decade, incomes for the top 1% of Americans have risen 40% compared with income declines for those in the middle (Stiglitz, 2011). This imbalance diminishes happiness and well-being, and stimulates dissatisfaction, conflict, and protests. As Gini indices rise, conflicts intensify. For example, recall the 2011 Occupy Wall Street protests against financial greed that swept the United States and spread to Europe and Asia (Bloomberg, 2011).

Surely there are other sources of happiness and well-being besides income that upsets us when it is unequally distributed. And surely there is more to life than running on the hedonic treadmill that gives only sporadic and temporary boosts. Indeed there is more. Other sources of happiness create internal well-being rather than polish its surface.

5.2.2 Relationships, Health, and Meaning

Beyond "buying things" and "keeping up with the Joneses," both of which are possible through economic growth, other more durable forms of happiness and

---

6. CIA (2014); data reflects income inequity from 2012 or earlier, depending on availability.

well-being can offset the short-term self-mindedness of our mental model. Psychologists, economists, and business leaders alike find that happiness and well-being depend more upon the quality of relationships, a sense of belonging, and feelings of significance or meaning (consistent with Eastern philosophies) than on "the quantity of goods we consume" (Daly, 2007), which reflects a Western influence.

Stefan Klein's perspectives bring neuroscience to bear. "A civic sense, social equality, and control over our own lives," he says, "constitute the magic triangle of well-being in society" (Klein, 2002). Others suggest that employment and job satisfaction contribute to our deepest happiness (Diener and Seligman, 2004). Surveys support these conclusions. A 2011 Gallup World Poll found that "a good job" with "work that is meaningful and done in the company of people we care about" is the number one determinant of happiness.[7]

In 2012, Gallup measured positive emotions in 143 countries with questions that reflected internal sources of happiness such as feeling respected, laughing, and learning or doing something interesting (Clifton, 2013). Latin American countries Paraguay, Venezuela, Panama, and Colombia had the highest "positive experience index" and Middle East countries such as Syria and Iraq, where civil strife is a daily occurrence, had the lowest.

As expected, incomes in most of these countries do not correlate with positive emotions. If they did, Panama, whose average monthly wage is $831, and Colombia, whose average monthly wage is $692 (relative to the world average of $1,480), would have low positive experience indices. The United States and the United Kingdom, with their high average monthly wages of over $3,000, would have ranked higher than 26th and 27th. On the other hand, Syrians have extremely low average monthly incomes of $362 *and* low positive emotions – an indication of national turmoil and living below the subsistence level.[8]

Personal experiences further underscore the nonmaterial aspects of happiness. In his classic *Man's Search for Meaning*, psychologist and concentration camp survivor Victor Frankl grasped the significance of having meaning in one's life. "Success, like happiness, cannot be pursued," he says, "it must ensue, and it only does so as the unintended side effect of one's dedication to a cause greater than oneself" (Frankl, 1985).

Physical fitness and mental health also have marked effects on well-being (Diener and Seligman, 2004). Medical advances have increased average life expectancy around the world. From an average of 48 years in 1955 and 65 years in 1998, global life expectancy should reach 73 in 2025; no country's average will be below 50 (WHO, 2014). Increased longevity with better health enriches

---

7. Mackey and Sisodia (2014) cited the Gallup Poll as described in Clifton (2011).
8. BBC (2012); Data in purchasing power parity $2009. Original statistics from International Labour Organization. Retrieved from <http://www.ilo.org/travail/areasofwork/wages-and-income/WCMS_142568/lang--en/index.htm>. Average wage in the United States was $3,263 and in the United Kingdom it was $3,065.

happiness and well-being. We will explore the downsides of these quality of life improvements in Chapter 6, namely that increased longevity brings the challenge of an aging population.

5.3 LONG-TERM/SHORT-TERM BALANCE

There is still an unanswered question regarding the two faces of happiness: Must we choose between them? We know that promoting personal gratification through material consumption harms the environment. We also know that there are other ways to feel satisfaction. Should we conclude then that personal gratification is verboten? Must we stop the buying that makes us feel good? Not entirely. We humans need immediate stimulation *as well as* long-term satisfaction (Offer, 2007).

In his seminal work, famed motivational researcher Ernest Dichter responds to this dilemma. He reminds us that "to want material things ... is a natural human desire, engendered partly by biological forces, the wish for security and protection, and reinforced by our contemporary culture" (Dichter, 1960). We cannot escape the pull of this desire. Instead, we must appreciate the dual reasons for consuming more than what we need to survive, namely to achieve instant gratification *and* to satisfy a deeper human need for long-term security. However, to adapt to a future that dictates less consumption, we must also pursue the nonmaterial sources of happiness and well-being: the feeling of belonging from relationships, the sense of significance from meaningful endeavors, and the contentment of being healthy. Yet, we must neither elevate immediate gratification above sustainable happiness nor focus on long-term happiness at the expense of short-term gratification. In other words, our coin of happiness and well-being must strike a balance between the two faces.

Achieving this balance is not simply a matter of choice or will. Because happiness is a fundamental element in our systems perspective of sustainability, we must take care with how we go about achieving it. As we will discover in Chapter 9, most actions in complex systems can have opposing effects. Short-term benefits may create long-term repercussions and long-term benefits can harm us in the short term. We have already noted that by emphasizing the personal gratification face of happiness, our mental model leads us to undesirable outcomes. Still, if we were to prevent this future harm by completely halting material consumption, our present economies would fail.

Having good health care exemplifies another case where short-term benefits can cause long-term damage. On the short-term side, good health care reduces infant mortality, increases longevity, and promotes well-being and happiness. In the long term, it enables population growth that places greater stress on food and water supplies, increases consumption and pollution, and depletes nonrenewable energy sources more rapidly. These consequences eventually diminish quality of life and reduce population through disease, hunger, and conflict – an unintended outcome of health care that was meant to promote well-being!

To further complicate matters, both faces of happiness – sustainable well-being and personal gratification – are so completely entwined with economic growth that it is difficult to sort out cause and effect. For instance, in nations whose robust economies promote personal gratification, lifestyles have become more sedentary and less healthy. Treatment of chronic disease from these lifestyles translates into negative dollars and cents for an economy. In the United States, health care costs totaled 16% of its GDP in 2007 (CDC, 2013). Health care has gone out of reach for some. Paradoxically, in this instance, two unintended long-term consequences of economic growth are poor health and inadequate health care, which diminish well-being.

The lesson from these cases is that we must not choose one path or the other, nor can we focus on one part of the issue. Our solutions must incorporate balance between present and future, and must view problems through a system lens. In other words, we should not completely *stop* efforts that will help us in the present. Instead we should modulate them and counter-balance their negative long-term effects on the rest of the system.

This discussion about blending the two faces of happiness has merely skimmed the surface of the short-term/long-term trade-offs required to achieve sustainability. We will discuss more of these dilemmas in subsequent chapters and ensure that proposed actions reach a compromise between present and future outcomes. We must temper our excesses but maintain our economies, nurture our health but reduce our population, and cultivate patience and appreciation for long-term satisfaction in life while enjoying some level of personal gratification. The next chapter moves us a step closer to understanding relationships among the three components of sustainability. It underscores the importance of population and pollution and brings them squarely into our field of view.

REFERENCES

BBC, 2012, March 29. Where are you on the global pay scale? BBC News Magazine. Retrieved from <http://www.bbc.com/news/magazine-17543356>.

Bloomberg, 2011, October 15. Protests spread as thousands gather in Europe, Asia. Reader Supported News. Retrieved from <http://readersupportednews.org/news-section2/316-20/7897-protests-spread-as-thousands-gather-in-europe-asia>.

CDC, 2013, October 23. Rising health care costs are unsustainable. Centers for Disease Control and Prevention. Retrieved from <http://www.cdc.gov/workplacehealthpromotion/businesscase/reasons/rising.html>.

CIA, 2014. The world factbook. Central Intelligence Agency. Retrieved from <https://www.cia.gov/library/publications/the-world-factbook/>.

Clifton, J., 2011. The Coming Jobs War. Gallup Press, New York.

Clifton, J., 2013, September 30. Syrians, Iraqis least positive worldwide: Latin Americans still most positive. Retrieved from <http://www.gallup.com/poll/164615/syrians-iraquis-least-positive-worldwide.aspx>.

Conference Board of Canada, 2013. World income inequality. How Canada performs. Retrieved from <http://www.conferenceboard.ca/hcp/hot-topics/worldinequality.aspx>.

Daly, H., 2007. Ecological Economics and Sustainable Development, Selected Essays of Herman Daly. Edward Elgar, Northampton, MA.

Dichter, E., 1960. The Strategy of Desire. Doubleday & Company. Garden City, New York.

Diener, E., Seligman, M., 2004. Toward an economy of well-being. Psychological Science in the Public Interest 5 (1), 1–31.

Edwards, A., 2010. Thriving Beyond Sustainability: Pathways to a Resilient Society. New Society Publishers, Gabriola Island, BC.

Frankena, W., 1973. Ethics, second ed. Prentice-Hall, Englewood Cliffs, NJ.

Frankl, V., 1985. Man's Search for Meaning. Washington Square Press, New York.

Frey, B., Stutzer, A., 2002. Happiness and Economics: How the Economy and Institutions Affect Human Well-being. Princeton University Press, Princeton.

Gore, A., 2013. The Future: Six Drivers of Global Change. Random House, New York.

Heinberg, R., 2011. The End of Growth: Adapting to Our New Economic Reality. New Society Publishers, Gabriola Island, BC.

Hernández-Murillo, R., Martinek, C., 2010, January. The dismal science tackles happiness data. The Regional Economist. Retrieved from <http://www.stlouisfed.org/publications/re/articles/?id=1860>.

Klein, S., 2002. In: Lehmann, S. (Ed.), The Science of Happiness: How Our Brains Make Us Happy – and What We Can Do to Get Happier. Da Capo Press, Cambridge, MA.

Mackey, J., Sisodia, R., 2014. Conscious Capitalism: Liberating the Heroic Spirit of Business. Harvard Business Review Press, Boston.

Mowday, R., 1991. Equity Theory Predictions of Behavior in Organizations. In: Steers, R., Porter, L. (Eds.), Motivation and Work Behavior. fifth ed. McGraw-Hill, New York, pp. 111–131.

Offer, A., 2007, November. A vision of prosperity: think-piece for the SDC Workshop Visions of Prosperity. Sustainable Development Commission, London.

Schmookler, A., 1993. The Illusion of Choice: How the Market Economy Shapes Our Destiny. State University of New York Press, New York.

Shaw, G.B., 2000/1903. Man and Superman. Penguin Books, London.

Smith, L., 2010. The World in 2050: Four Forces Shaping Civilization's Northern Future. Dutton, New York.

Stiglitz, J., 2011, May. Inequality: of the 1%, by the 1%, for the 1%. Vanity Fair. Retrieved from <http://www.vanityfair.com/society/features/2011/05/top-one-percent-201105?currentPage=all>.

WHO, 2014. World health report. 50 Facts: Global Health Situation and Trends 1955–2025. World Health Organization. Retrieved from <http://www.who.int/whr/1998/media_centre/50facts/en/#>.

Chapter 6

The Bubble Bursts: Population and Pollution Become Our Concern

Throughout the past two centuries economic growth has translated to an increased capacity to support more humans with Earth's available resources. ...Human population growth can be seen as an indication of our success as a species. But now ... higher population levels pose an enormous vulnerability.

– Richard Heinberg (2011).

The earth does not belong to man, man belongs to the earth. All things are connected like the blood that unites us all. Man did not weave the web of life, he is merely a strand in it. Whatever he does to the web, he does to himself.

– Chief Seattle, 1854[1]

This chapter and the next two are all about altering our mental model – about bursting our bubble of beliefs on the prickly ground of reality as in Fig. 6.1. Like a toddler who cries when her soap bubble pops before she can touch it, we will no doubt feel disappointment when a deeper awareness of our world punctures our convictions.

Part of this awareness may already live in the back of our minds. Perhaps we sense there is a limit to the natural resources that make our economy and our pocketbooks grow. We must have an inkling that all this economic growth has significant repercussions and that there are penalties for our complaisance. Perhaps we wonder whether economic growth has enabled population growth and contributed to pollution. Perhaps there is something to the global warming that everyone is talking about. For most of us, all these "perhapses" are irksome flies buzzing around our heads; we just swat them away. However, judging from copious literature on the topic, uneasiness about constraints and consequences is intensifying.

---

1. Legend says that Chief Seattle (Sealth) of the Duwamish and Suquamish Native American tribes spoke these words in 1854. Chief Seattle quoted in Baring and Cashford (1992).

K.L. Higgins: Economic Growth and Sustainability. http://dx.doi.org/10.1016/B978-0-12-802204-7.00006-2

FIGURE 6.1   Bubble about to burst.

This chapter recalibrates our mental model. It examines two major factors that we have pushed outside its boundaries: (1) population and (2) pollution.[2] To integrate these realities and to capture the interdependencies and dynamics of the real world, we rely on systems thinking. Recall that systems thinking does not require quantified data, but does incorporate trends that the data describes. Thus, when you view the many charts and graphs in this chapter, do not fixate on the numbers. Concentrate instead on how rapidly the elements of interest move with passing time. Compare their ups and downs. And, rather than questioning whether one individual or one country fits these trends, look at global aggregates and imagine the powerful momentum behind them. These trends will become a part of the system diagram in Chapter 8.

6.1   POPULATION GROWTH

Since the first bands of humans roamed the African plains, the number of people on Earth has escalated. Societies have spread to every habitable corner. Initially, population growth was tempered by disease and harsh living conditions; it took a thousand generations for world population to increase from three million to ten million when agriculture reached the Fertile Crescent of Western Asia.[3] Although this 10 million mark was huge at the time, it is smaller than Bejing today.

From an estimated 370 million people who remained after famine and Black Death took their tolls in the fourteenth century, the number has soared. Spurred by productivity and economic advances in the Industrial Age, world population grew to one billion in 1804 and tripled to three billion in 1959. By 1999, it had doubled again to six billion. Nine billion people will most likely coexist by 2042.[4]

This staggering number seems so inconceivable that anxiety about whether Earth can support all these people is mounting. Yet, concerns about population growth are millennia old. Since Confucius in the sixth century BCE, Plato and Aristotle in the fifth century BCE, and Kautilya in 300 BCE, philosophers have warned that excessive population growth will degrade quality of life

---

2. To reduce complexity, we will not explicitly include other critical social factors such as poverty, disease, and education in this section. They will appear in later chapters.
3. Hoggan (2010); the time frame for this increase was from 35,000 BCE to 10,000 BCE.
4. Population data from United Nations (1999) and from US Census Bureau (2013a).

(Neurath, 1994). Centuries later, in the 1700s, British economist Thomas Malthus cautioned that population would outgrow food supply (Malthus, 1988). In retrospect, these warnings seem trivial, especially since technology and innovation found a way to overcome the challenges of food supply and well-being. Are things different today or will we again accommodate our growing numbers? Let us consider some statistics that are unique to our times and make us think about our future.

6.1.1   Population Statistics

Figure 6.2 charts population change for a century. Although the population growth rate has declined since 1963, it is not zero: population still rises.

Aggregated statistics such as these can be misleading. First, projected global population reflects most likely scenarios based on history and current trends. Projections do not account for potential environmental or social variables such as AIDS and lack of water, or couples who choose to have fewer children (Brown et al., 2000).

Second, a big picture perspective hides important details. For example, it considers individuals as world citizens and nations as world constituents who share the same problems. Although this view works for many global issues, it glosses over important differences related to population. We are, in fact, a world of many nations separated by semirigid boundaries. This segregation prevents a free flow of people and creates specific population demographics and unique problems for each nation.

Coincidently, regions in which population is rising the fastest also have the least water and arable land to support their growing numbers. Between 2015 and 2040, Africa's population should increase by 65%, the near East's population by 34%, and India's population by 26%. These dry areas will be first to feel the stress of hunger, thirst, and disease. Alternatively, population will rise only 21% in North America and will remain stable or decline in Eastern Europe, Western Europe, and China (US Census Bureau, 2013b).

6.1.2   Fertility Rates

Between 2015 and 2040, global fertility rate will drop from 2.4 to 2.2 births per woman (US Census Bureau, 2013b). Depending on culture, religion, laws, and affluence, individual nations have different fertility rates and thus experience different population issues. In general, in countries with fertility rates higher than the replacement rate of 2.1 children per woman, populations increase; in countries with lower fertility rates, populations decline.

Figure 6.3 contrasts fertility rates in 1991 with those in 2011 for nations whose rates are among the highest and lowest in the world. Of the 18 listed, women in certain African and Middle Eastern countries have the most children; Niger tops the list at an average of nearly eight per woman. Alternatively, in 11 countries,

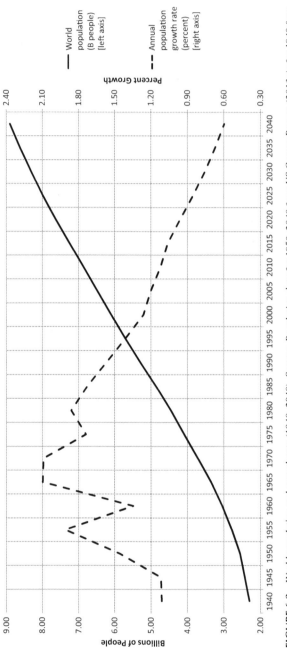

FIGURE 6.2 World population and growth rate (1940–2040). *Source: Population data for 1950–2040 from US Census Bureau (2013a); for 1940 from US Census Bureau (2013c) (data interpolated for 1945 estimate).*

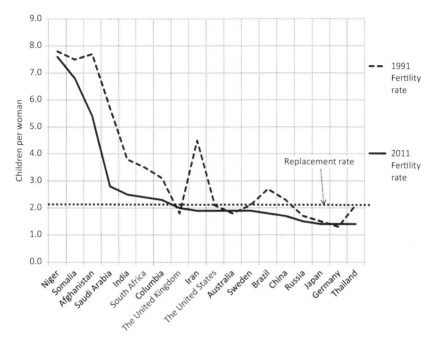

FIGURE 6.3 Fertility rates of selected countries (1991 and 2011). *Source: Fertility rates from The World Bank (2014).*

including the United Kingdom, Iran, the United States, China, and Japan, fertility rates are low and populations are relatively constant or are dropping.

Countries with population growth experience two conditions. On one hand, more people means more pollution and a greater need for food, water, and energy as well as more children to educate and eventually employ. On the other hand, more people implies that there are more consumers to grow the economy and more young employees to replace retirees, augment a country's tax base, and support the elderly.

Countries where population is decreasing face the challenge of increased median age, which burdens younger generations and threatens economies. Here, the number of older people increases and the number of younger people decreases. Because global population growth rate is tapering off, more nations will share this burden.

6.1.3 Birth and Death Rates

In the 50 years between 1965 and 2014, population growth rate dropped from 2.1% to about 1%.[5] Even with this downward trend (see Fig. 6.2), we might ask why population is not declining as rapidly as the decrease in fertility rates

5. Population growth rate and birth rates for 1965 from US Census Bureau (2013a); for 2014 from CIA (2014).

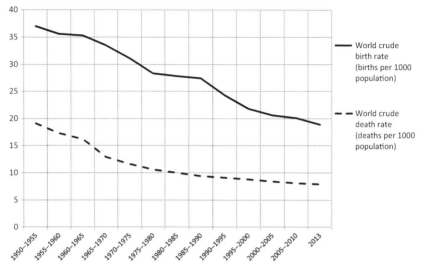

FIGURE 6.4 World birth and death rates (1950–2013). *Source: World crude birth rate and birth rate 1950–2010 from United Nations (2013); for 2013 retrieved from <http://www.indexmundi. com/world/birth_rate.html> and <http://www.indexmundi.com/world/death_rate.html>.*

in Fig. 6.3 suggests. William Ryerson, founder of the Population Media Center, states that surging population growth rates primarily result from "declining death rates – the result of widespread vaccination programs and other public health measures" (Ryerson, 2010).

To visualize this mortality phenomenon, notice that the lines representing birth and death rates in Fig. 6.4 draw closer together as time passes. Between 1950 and 2013, the number of deaths per 1000 population decreased from 19 to 7.9 (nearly a 60% dip in about 60 years). On the other hand, births per 1000 dropped 50% from 37 to 18.9, or 10% less than the change in death rate.

Just as with fertility rates, birth and death rates differ by country. South Africa's estimated 2014 death rate of 17.5 per 1000 ranked highest; Qatar's 1.5 was lowest (CIA, 2014). African countries also had high birth rates; Niger's estimated 2014 birth rate per 1000 population is 46.1. Japan's birth rate of 8.1 is near the bottom and is below its death rate (CIA, 2014). Because overall death rate has declined faster than birth rate, population has increased.

6.1.4 Median Age, the Elderly Dependency Ratio, and the Economy

World population demographics are changing for many reasons. Two primary causes are acceptance of family planning, which has reduced fertility and birth rates, *and* better health care, which has lowered death rates. Thus, there are more people in the world but median age has changed. It hovered around

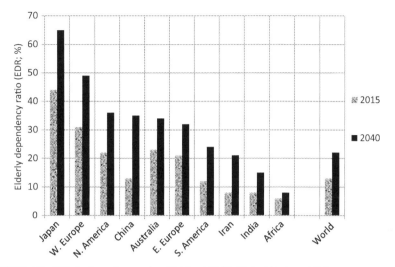

FIGURE 6.5 Elderly dependency ratios (2015 and 2040) (United Nations, 2013).

24 years for half a century and only began to rise in 2000.[6] In 2014, estimated median age was 29.4 years (CIA, 2014); it is projected at 32 years in 2025 and 36 years in 2050.

Increased median age has several systemic implications. One is that aging detracts from well-being when the elderly have more health issues. Another is that aging hinders the economy in many countries. When the number of elderly grows faster than the number of working age people (consistent with increasing median age), there are fewer young workers to support more elderly. Furthermore, a decreased number of young workers can create a shortage of employees. These conditions hurt the global economy.

Different median ages in individual nations emphasize that nations face different issues. In 2010, for example, median age in more economically developed regions was age 40; in least developed countries it was only age 19 (United Nations, 2013).

One metric, called the "elderly dependency ratio (EDR)", quantifies the effects of median age. It reflects the proportion of "elderly" people (age 65 and up) to "working age" individuals (between age 15 and 64). When EDR rises, the number of elderly is growing faster than the number of working age people. Figure 6.5 projects that world EDR will increase as the number of working age people per elder drops nearly in half from 7.7 to 4.5. Of all countries, Japan will be hardest hit; between 2015 and 2040, the number of working age per elder will drop from 2.3 to 1.6. On the other hand, in Africa during the same period,

---

6. Median age statistics from CIA (2001) and United Nations (2013).

the number of working age per elder will decline less dramatically from 16.6 to 12.5; EDRs in China and Iran will increase the fastest – both by at least 60%.[7]

This maze of statistics points out that when the rate of population growth slows, median age increases and economies must support more elderly. Estimates of public spending for the elderly illustrate these trends. In 1995, for example, the United States, the United Kingdom, and Japan spent 10–12% of their GDPs on pensions and health care benefits. By 2030, these numbers should hit 15–23%.[8]

Earth's population is growing like algae in a pond and is aging more rapidly than ever. Unlike Aristotle or Malthus who could not conceive of the miraculous technology that saved the day for us, we now face the sobering fact that population is an issue and that there are actual limits to Earth's capacity to sustain these massive numbers, even with technology. We explore these limits in Chapter 7.

6.2 INCREASED POLLUTION

Apprehension about environmental issues is more recent than concern about population. Since the Industrial Age began in the 1760s, technology has multiplied man's damage to the environment – to air and land and water. Until these effects became too large to ignore, society emphasized economic growth without regard for the environment. For example, it was not until 1952 – after the "Great London Smog" had killed 12,000 people – that Britain introduced its Clean Air Act.[9] Rather than wait until extreme disaster strikes, let us reconsider our disregard for the environment.

Natural systems produce waste. Whether we are talking about the human body or about the economy, all convert raw materials into various forms of waste. "Pollution" is a catch-all term for these by-products. This section concentrates on five primary pollutants that significantly affect sustainability: (1) greenhouse gas (GHG), (2) air pollution, (3) municipal solid waste, (4) radioactive waste, and (5) industrial, agricultural, and human wastes.

Our mental model neglects the systemic and detrimental effects of these pollutants. In truth, they feed back to impair the economy, environment, and society. For example, a recent report from the Blacksmith Institute documents the repercussions of pollution from sources such as industrial plants, mining facilities, electronic waste, and nuclear disaster sites. Their report ties environmental factors to nearly a quarter of deaths in the developing world and about a fifth of

---

7. Between 2015 and 2040 working age people per elder in China drops from 7.7 to 2.8 (ratio increases from 13 to 35) and from 12.5 to 4.8 per elder in Iran (ratio increases from 8 to 21). See United Nations (2013).

8. CIA (2001). The United States and the United Kingdom spent 10.5% and Japan spent 11.5% of their GDP in 1995; official projections for 2030 in the United States, the United Kingdom, and Japan are 15.5%, 17.0%, and 23.1% of GDP, respectively.

9. See J. Rosenberg, The great smog of 1952. Retrieved from <http://history1900s.about.com/od/1950s/qt/grsmog.htm>.

cancer incidents for the entire world (Blacksmith Institute/Green Cross, 2013). The following discussions describe how major sources of pollution affect us.

6.2.1 Greenhouse Gas

To generate the lion's share of the energy used in homes, industries, and agriculture, we devour huge amounts of fossil fuels that produce GHG. Added to these sources, the burgeoning livestock industry accounted for 14–18% of total GHG in 2006 as world diets trended toward eating more meat.[10] Thus, a growing economy and a growing population consume more energy and produce excessive GHG.

Greenhouse gases that concern us most are carbon dioxide (CO_2), methane, nitrous oxide, and fluorinated gases. CO_2, most of which comes from fossil fuel consumption, accounts for over three-quarters of these (IPCC, 2007). Concentration of CO_2 has increased since 1750 "due to human activity" (IPCC, 2013) and has accelerated over the past several decades.[11]

But so what? Why should we care about invisible particles in the air? In moderation, greenhouse gases in the atmosphere are a good thing. They absorb and emit thermal energy from the sun and create a sort of cocoon that keeps warmth inside and makes our planet habitable. However, in excess, GHG changes the composition of this cosy little cocoon and causes it to trap too much heat.

Although debates rage on, hundreds of scientists around the world believe that high levels of GHG are creating a condition called "global warming." They predict weather extremes and other ill effects of unprecedented high temperatures. Earth's surface temperature has in fact increased in each of the last three decades – more than in any decade since 1850. Researchers conclude that "in the Northern Hemisphere, 1983–2012 was likely the warmest 30-year period of the last 1400 years."[12]

To be fair, greenhouse gas is not the only cause of global warming. Other natural contributors include what is called the "Milankovitch cycle," that is, periodic changes to Earth's axial rotation and its orbit around the Sun.[13] Nevertheless, this book aligns with preponderant evidence that human-generated greenhouse gases have dramatic effects on climate; the integrated system diagram includes these effects and keeps planetary action outside the system boundary.

---

10. FAO (2006), xxi and 10; Herzog (2009).

11. Butler and Montzka (2013). From 1.4 parts per million (ppm) per year between 1979 and 1995, the average growth rate climbed to 1.9 ppm per year after 1995.

12. IPCC (2013). 800 researchers from 39 countries concluded that the Earth's climate is warming and that "human influence on the climate system is clear."

13. For a description of the Milankovich cycle, see <http://ossfoundation.us/projects/environment/global-warming/milankovitch-cycles>. These natural cycles are thought to occur about every 100,000 years and include brief warm periods followed by an ice age.

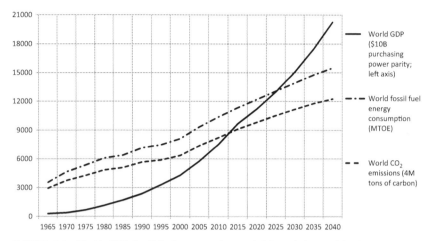

FIGURE 6.6 World GDP, fossil fuel consumption, and CO_2 emissions (1965–2040). *Source: GDP from 1980 to 2015: International Monetary Fund (2013). For 1970 and 1975: International Monetary Fund (1999). GDP Growth for 2020 projected at 3% per year from Pricewaterhouse-Coopers (2013). Energy consumption and CO_2 emissions 1965–2010 from BP (2014). Projections 2015–2040 using annual percent increase for CO_2 emissions from EIA (2013).*

Figure 6.6 illustrates that GDP, fossil fuel consumption, and the CO_2 emissions that contribute to global warming are simultaneously rising and will continue to do so. A recent study predicts that GHG levels may climb by as much as a third in the next 20 years. Although shale gas proponents argue that using natural gas will cut emissions, world dependence on coal (a huge source of CO_2) overshadows these reductions.[14]

Recent history validates concerns about growing CO_2 concentration. From about 320 parts per million in 1964, it reached a record high of 400 parts per million in May 2013 and again in March 2014. These levels are evoking cries for action.[15]

Now comes the question: How much greenhouse gas is too much? Our experience here is nonexistent. However, scientific data can inform our predictions. Figure 6.7 superimposes rising CO_2 levels onto warming trends and estimated danger thresholds. Note that if temperatures increase even 0.6 °C above 2010 levels, Earth will touch the zone of substantial to severe risk.[16]

If we reach these thresholds, water levels will rise. High temperatures will devastate coral reefs, glaciers, small island states, forests, and unique ecosystems, and will increase the frequency, intensity, and consequences of heat-waves,

---

14. Harvey and Macalister (2014). Study conducted by BP.
15. BBC (2013) and CO_2 now, 2014. In May 2013 and again in March 2014, CO_2 concentration reached 400 parts per million at Mauna Loa Observatory in Hawaii.
16. Smith et al. (2009). Surface temperature rose 0.4 °C between 1990 and 2010. Estimated increase between 2010 and 2040 exceeds 0.6 °C and ranges from 0.7 °C to 6.0 °C.

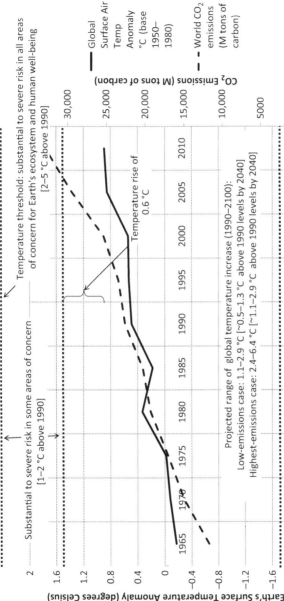

FIGURE 6.7 Global surface air temperature anomalies and CO_2 emissions (1965–2040). *Source: Global surface air temperature anomalies and historic references from NASA (2013). (Global mean temperature is ~14 °C; to estimate an absolute global mean temperature, add the temperature anomaly to mean temperature). CO_2 emissions 1965–2010 from BP (2014); projections 2015–2040 from EIA (2013) using annual percent projected increase. Areas of risk and repercussions from Smith et al. (2009).*

FIGURE 6.8   Time elapse calving of an Alaskan glacier.

droughts, floods, and wildfires across large regions. Although the question of whether Arctic and Antarctic ice caps are melting is controversial, satellite photos from a recent study show accelerated Arctic ice loss since 1992, particularly in Greenland. Melting ice, caused in part by warming oceans and thawing mountain glaciers, has been "responsible for a fifth of the global rise in sea levels since 1992, 11 millimeters in all" (Quaile, 2013). Fig. 6.8 brings these statistics into the real world. It illustrates a typical glacier melting process when massive hunks of ice suddenly sheer off and fall into the ocean.

Even today as CO_2 increases, Earth is experiencing weather extremes that affect our well-being and our economies. In the United States, for example, more intense hurricanes and coastal storms have resulted in an estimated annual property loss of $35 billion and have cost corn and wheat farmers tens of billions of dollars annually; heat-wave-driven demand for electricity has cost consumers up to $12 billion a year (Begley, 2014). Furthermore, severe drought in many parts of the world (see Chapter 7) has already harmed both farmers and individual citizens.

Unfortunately, damage begets more damage. NASA is currently researching whether rising temperatures create conditions that accelerate climate change. Scientists fear that more CO_2 and methane will be released when warmer temperatures thaw Arctic permafrost and decompose organic materials; this reaction will further increase temperatures and create a vicious circle of heat $\leftrightarrow$ greenhouse gas (Healy, 2013). Climate scientist Amanda Staudt's research points out another subtle implication of global warming: "Every degree Celsius brings about 6% more lightening" which triggers more fires (cited by B. McKibben, 2010) which, of course, release more greenhouse gases and other pollutants.

6.2.2 Air Pollution

Most major world cities are plagued by air pollution. Its noxious particulates include sulfur dioxide, nitrogen dioxide, carbon monoxide, ozone, smog, and hydrocarbons such as methane. Internal combustion engines, forest fires, wood, and coal burning stoves, rubbish incineration, and dust number among its sources. In addition to being a major component of GHG, CO_2 contributes to air pollution; when combined with moisture, it creates acid rain that damages whatever it touches.

Cities with the highest concentration of fine particulates include Delhi, Karachi, Abu Dhabi, Dakar, and Bejing. Notably, many of these cities are the world's most populous. Polluted air affects children, the elderly, and the frail the most, and causes stroke, heart disease, respiratory infection, and lung cancer for many others. The World Health Organization reported that in 2012, one in eight deaths worldwide was related to air pollution. Air pollution also causes rust, damages stone buildings, decomposes nylon, and kills plants.[17]

6.2.3 Municipal Solid Waste

Another category of waste, called "municipal solid waste (MSW)," is the garbage that humans dispose of every day (excluding industrial, agricultural, and human wastes). Its growing presence reflects the abandon with which we discard the remnants of our lives. In addition to the space it occupies and the money it costs to dispose of or to recycle, MSW contaminants leach into water sources, generate air pollution when they are burned, and produce almost 5% of all greenhouse gas (Hoornweg and Bhada-Tata, 2012).

As one would expect, Fig. 6.9 shows that the most affluent nations (most OECD countries) generate more waste per capita than elsewhere. In fact, high-income individuals generate as much as 31 pounds of waste a day compared with low-income individuals who generate as little as a fifth of a pound. Of note is that the densely populated East Asia Pacific region (including China and Thailand) will have the most *total* waste by 2025; between 2012 and 2025 its waste will increase over 150% compared with 11% in OECD countries. Thus, with accelerating economic growth in newly emerging, highly populated nations, we can expect more waste and its resultant environmental and health issues.

A picture can imprint the magnitude of this problem on our brains. The Indonesian teenager in Fig. 6.10 collects plastic from one of Jakarta's polluted rivers. Water pollution in Indonesia is severe. The world's worst polluted river, the Citarum, supplies most of the water for Jakarta's people, farms, and

---

17. Data on air pollution in this section from WHO (2014). Particulates are categorized by size. This data refers to particulate matter that is 10 microns or less, also PM_{10}.

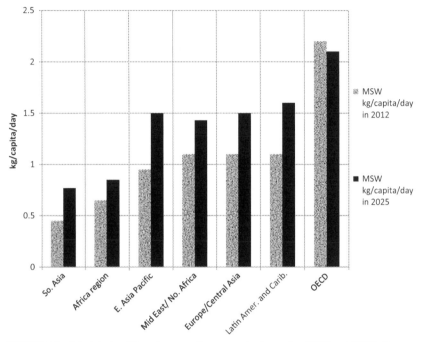

FIGURE 6.9 Municipal solid waste generation per capita by region (2012 and 2025). *Source: Data from Hoornweg and Bhada-Tata (2012). The Organisation for Economic Co-operation and Development (OECD) includes the United States, the United Kingdom, Japan, South Korea, Australia, and several European countries.*

FIGURE 6.10 Water pollution in Indonesia. *Source: A scavenger collects valuable goods at a polluted river near Pluit dam in Jakarta June 5, 2009. Photo Credit: REUTERS/Beawiharta.*

FIGURE 6.11 E-waste dumpsite in Ghana. *Source: Photo reproduced with the permission of Blacksmith Institute, <www.blacksmithinstitute.org>.*

factories; about nine million Indonesians come into contact with it. In addition to garbage, the Citarum contains over a thousand times the US environmental standards for lead and extreme concentrations of aluminum, manganese, and iron (Blacksmith Institute/Green Cross, 2013).

Unfortunately, cleaning up MSW can cause more problems than the waste itself, as exemplified by the e-waste dumpsite in Agbogbloshie, Ghana in Fig. 6.11. Ghana imports computers, microwaves, and other e-waste products to recycle. Open air burning of these materials generates smoke laden with heavy metals that find their way into the soil, water, and air, and put some 40,000 of Ghana's citizens at severe risk.

6.2.4   Radioactive Waste

Radioactive waste is unique. Operational nuclear power plants emit negligible GHG into the atmosphere relative to carbon-based fuels, thus immediate effects of nuclear energy production on the environment are minimal. In addition, unlike other forms of energy, costs to handle radioactive waste are already incorporated in the cost of electricity derived from nuclear plants.

The tradeoff with nuclear energy is that its physical waste poses severe health hazards and must be disposed of according to strict laws and regulations. Categories of radioactive waste depend on their potential for harm. Low-level waste, which includes medical trash and material from dismantled nuclear sites, can be buried in shallow land. Intermediate levels (e.g., from decommissioning

FIGURE 6.12   Nuclear power station and radioactive waste storage. *Source: All photos courtesy of US Nuclear Regulatory Commission. Clockwise from upper left: Beaver Valley Nuclear Power Station, Shippingport, PA; low-level radioactive waste storage site , Hanford, WA; on-site spent fuel storage pool, San Onofre, CA; inside the high-level radioactive waste storage site, Yucca Mountain, NV.*

nuclear reactors) are temporarily held in recycling facilities, then permanently stored underground.[18]

Spent uranium from nuclear reactors is the most dangerous category; this high-level waste contains 90% of all radioactivity. It is initially housed in specially designed pools of water at reactor sites. When wet storage areas reach capacity, the spent fuel is permanently stored elsewhere. After a few years, it can be placed in aboveground metal or concrete storage casks, or buried deep in clay, rock salt, or granite underground facilities. The radioactivity in high-level waste takes at least 1000 years to reach safe levels (World Nuclear Association, 2013). Fig. 6.12 shows a nuclear power plant and various waste storage sites.

Unpredictable events and accidents are primary instigators of pollution from radioactive waste. The March 2011 magnitude 9.0 earthquake and tsunami in Japan devastated the Fukushima Daini nuclear plant and released 900,000 terabecquerels of radioactive materials. An estimated 100 metric tons of radioactive water leaked from the plant's holding tank into the ocean. Besieged by

---

18. Maitre and Boselli (2009) indicate that low-level waste accounts for 90% of all waste by volume, but contains less than 1% of the radioactivity. Intermediate-level waste makes up 4% of all waste and 8% of radioactivity.

continuing leakage and cleanup costs estimated at $40 billion, Japan took all 50 of its nuclear reactors off line in 2013.[19] Global warming trends also pose an indirect, but serious threat to nuclear waste storage. For example, thawing of the permafrost in Russia's Arctic region (where high-level radioactive waste is stored and nuclear weapons were tested) could expose the world to its hazards (RIA Novosti, 2014).

Worldwide nuclear power generating capacity is an indicator of radioactive waste. Between 1970 and 1990, capacity increased nearly 20-fold;[20] since then, growth has been modest.[21] The challenge of storing radioactive waste will increase as countries build more nuclear reactors to keep up with rising population and energy consumption.

6.2.5   Industrial, Agricultural, and Human Wastes

Industrial, agricultural, and human wastes pollute water, land, and air in similar ways. We have already described the effects of GHG and air pollution, some of which come from industry, agriculture, and humans. This section describes water and land pollution from these sources.

In addition to GHG and other air-borne particulates, industrial wastes release toxic heavy metals, ammonia, cyanide, and other hazardous chemicals into streams, rivers, and oceans, which then find their way into the ground. According to Blacksmith Institute's 2012 list of the most toxic industries, lead-acid battery recycling, lead smelting, mining, and tannery operations ranked highest. These and other contaminants cause cancer, kidney and lung disease, and neurological damage (Blacksmith Institute/Green Cross, 2012).

In addition to the air-borne methane and ammonia from livestock manure, erosion and dust from cultivation, as well as contaminants from pesticides and fertilizers, harm the environment. Water pollution devastates our ecosystems. It triggers algae blooms in coastal waters and threatens fish and wildlife, depleting oxygen and creating "dead zones" where few fish or wildlife can survive (Life of Earth, 2014).

Household and human waste multiplies with growing population. Lack of proper sanitation and inadequate sewage treatment spread cholera, typhoid, and hepatitis A. A case in point is the Ganges River Basin that sustains 43% of India's huge population and provides a quarter of their water. In Varanasi, one of India's seven sacred cities, much of the "sewage and industrial waste flows untreated into the river, alongside religious bathing" (Das and Tamminga, 2012). One might encounter floating human and animal bodies after certain religious

---

19. CNN (2014). Terabecquerel is a metric associated with radioactivity.
20. Capacities are estimated from Figure 1 in Sokolov and McDonald (2006), derived from International Atomic Energy Agency Energy, Electricity and Nuclear Power Estimates, July 2005. Capacity grew from a negligible 17 Gigawatts of electrical power (GWe) in 1970 to about 330 GWe in 1990.
21. IAEA (2013). Capacity was about 373 GWe in 2013.

purification ceremonies. Another example reinforces this sanitation problem; in 2013, less than two out of three people in the world had access to a toilet, while six out of seven had access to a mobile phone.[22] Inadequate sanitation and lack of sewage treatment in many parts of the world contribute to human misery. The World Health Organization reports that "half the developing world suffers from one of the six diseases associated with poor water supply and sanitation" (Brown et al., 2000).

6.3 INTERACTION AMONG POPULATION, ECONOMY, AND ENVIRONMENT

As world population grows, we might wonder what will happen in places that become overcrowded. We know from research (and possibly from personal experience) that when a casual acquaintance or a stranger gets too close, we become uncomfortable. In fact, neuroscientists have recently found that invasion of personal space triggers anger and fear in our brains.[23]

Our response to crowded conditions demonstrates the powerful emotions that such a teeming reality might provoke. Imagine a rush hour crowd in Sao Paulo, Brazil that pushes to enter a subway train; bodies are packed so tightly that no one can make an independent move. Or picture a train departing the Dhaka airport in Bangladesh in which passengers are not only riding cheek by jowl, but are also stacked on the top and hanging like ornaments off the sides. Even recreation can be uncomfortable. Fig. 6.13 shows a pool in China so overcrowded that it is impossible to move.

Now, close your eyes and transport yourself into the midst of these throngs. Imagine the pressure, pokes, smells, and body heat of strangers. What are your emotions – anxiety, fear, claustrophobia, panic? Although these musings are unsettling, invasion of personal space is but a minor implication of population growth.

What are the more harmful effects of so many people on the economy and the environment? Fig. 6.14 combines growth trends for population, fossil fuel consumption, and CO_2 emissions that we charted earlier, and adds trends for MSW and nuclear energy consumption. (Nuclear energy consumption substitutes for radioactive waste that increases with the amount of energy generated.)[24] Without doing a statistical analysis, we can deduce from the upward slope of all lines that there is a positive relationship among population, economy, and environment.

---

22. Wang (2013). Data from a 2013 United Nations study.
23. Kennedy et al. (2009). Comparison between individuals with normal amygdala and those with damaged amygdala noted different responses. Amygdala comprises two almond-shaped regions in the brain that link to strong negative emotions.
24. There are an estimated 270,000 tons of used fuel in storage; as of 2013, about 12,000 tons a year add to this accumulation. See World Nuclear Association (2013).

FIGURE 6.13 Crowded pool in Sichuan province, China. *Source: Visitors crowd an artificial wave pool at a tourist resort to escape the summer heat in Daying county of Suining, Sichuan province, July 27, 2013. Photo Credit: REUTERS/China Daily; Image ID: RTX1226J.*

Specific causes and effects are more complex and are difficult to infer from these trends. For example, because GDP is rising faster than population, we would expect that the average standard of living (measured by GDP per person) has increased. Post World War II, this observation is accurate; today, GDP is not entirely dependent on population.[25] Although more population increases the pool of consumers and the productive workforce, part of GDP growth comes from efficiencies and innovation brought about by technology (e.g., introduction of the computer in the 1970s), by manufacture of new and different types of products, by population demographics (particularly median age and the elderly dependency ratio described earlier), and by education levels that facilitate technology advances.

What we do not see with these global upward trends is what happens to population when GDP declines for an extended period. In fact we have no relevant history: Both GDP and population have continuously increased since the 1400s (DeLong, 1998). We do have anecdotal evidence of poverty and disease during periods such as the Great Depression in the 1930s, thus we will conclude that extended GDP decline decreases standard of living and diminishes population. We will also conclude that if population drops substantially, GDP would decrease but probably not as quickly.

---

25. DeLong (1998). From the early nineteenth century until WWII, there was a significant correlation between the rate of population growth and GDP per capita.

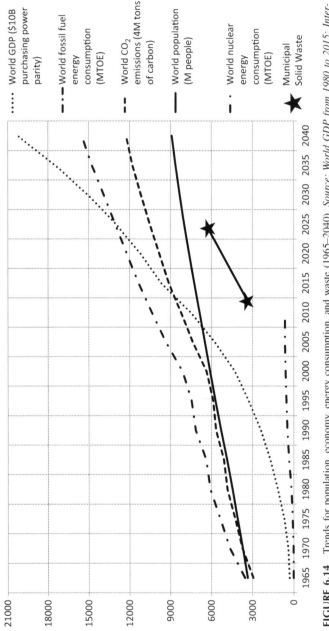

FIGURE 6.14 Trends for population, economy, energy consumption, and waste (1965–2040). *Source: World GDP from 1980 to 2015: International Monetary Fund (2013); for 1970 and 1975: International Monetary Fund (1999). GDP Growth for 2020 projected at 3% per year from PricewaterhouseCoopers (2013). Fossil fuel and nuclear energy consumption, CO_2 emissions 1965–2010 from BP (2014); projections 2015–2040 from EIA (2013) using annual percent projected increase for CO_2. MSW for 2012 and 2025 from Hoornweg and Bhada-Tata (2012).*

Riding the coat-tails of population and economic growth are increased energy consumption and pollution. As expected, fossil fuel consumption and CO_2 emissions are rising at about the same rate; both are increasing faster than population. This disparity is likely a result of economic boom in developing nations and their increased dependence on fossil fuels. The figure also shows that daily generation of MSW is projected to rise faster than population growth, indicating that economic growth *plus* population growth allows more people to buy and discard more *stuff*.

Thus, instead of considering pollution as someone else's problem or population as a solution to aging issues or an impetus to economic growth, we must revise our world view to incorporate their tight interdependence. This new view emerges as a system diagram in Chapter 8 and incorporates what we have learned so far: Population growth affects the environment and the economy (pollution and GDP); economic growth affects the environment and society (pollution and well-being); population decline creates social dilemmas (median age); and pollution affects society and the economy (well-being, land and water, and cleanup costs). We will also be able to integrate system delays that, for example, reflect the momentum behind population growth[26] and the accumulation of environmental damage.

Acknowledging the consequences of population growth and increasing pollution must forever change how we view the world. No longer can we push them aside, for they have already pricked the shimmery bubble that contains our mental model. When we integrate them into our worldview – into our system diagram – we will see the potential for disastrous outcomes.

It must be obvious from this chapter and from the previous two that the system that governs sustainability is extremely interdependent and more complicated than our mental model suggests. Understanding this system requires an integrated view of its elements, an ability to identify present behavior, and the foresight to predict the future implications of this behavior. It calls for a blending of disparities and a spirit of compromise. And, as we will see in Chapter 7, this long-term view recognizes that resources are finite and that growth is not everlasting.

---

26. If all women in the world had only the replacement number of children from today on, it would still take 60 years to achieve zero population growth (ZPG). During these 60 years, more women will reach childbearing age and will continue to have children. See Gardner and Stern (1996).

REFERENCES

Baring, A., Cashford, J., 1992. The Myth of the Goddess: Evolution of an Image. Penguin, London.

BBC, 2013, May 11. Scientists call for action to tackle CO_2 levels. BBC News Science & Environment. Retrieved from <http://www.bbc.co.uk/news/science-environment-22491491>.

Begley, S., 2014. U.S. to face multibillion-dollar bill from climate change: report. Reuters. Retrieved from <http://news.yahoo.com/u-face-multibillion-dollar-bill-climate-change-report-041353090.html>.

Blacksmith Institute/Green Cross, 2012. The world's worst pollution problems: assessing health risks at hazardous waste sites. Retrieved from <http://www.worstpolluted.org/files/FileUpload/files/2012%20WorstPolluted.pdf>.

Blacksmith Institute/Green Cross, 2013. The world's worst 2013: the top ten toxic threats: cleanup, progress and ongoing challenges. Retrieved from <http://www.worstpolluted.org/docs/TopTenThreats2013.pdf>.

BP, 2014, June. BP statistical review of world energy 2014. Retrieved from <http://www.bp.com/en/global/corporate/about-bp/energy-economics/statistical-review-of-world-energy.html>.

Brown, L., Gardner, G., Halweil, B., 2000. Beyond Malthus: Nineteen Dimensions of the Population Challenge. Earthscan, London.

Butler, J., Montzka, S., 2013. The NOAA Annual Greenhouse Gas Index (AGGI). National Oceanic and Atmospheric Administration. Retrieved from <http://www.esrl.noaa.gov/gmd/ccgg/aggi.html>.

CIA, 2001, July. Long-term global demographic trends: reshaping the geopolitical landscape. Central Intelligence Agency. Retrieved from <https://www.cia.gov/library/reports/general-reports-1/Demo_Trends_For_Web.pdf>.

CIA, 2014. The world factbook. Central Intelligence Agency. Retrieved from <https://www.cia.gov/library/publications/the-world-factbook/>.

CNN, 2014, February 20. Japan earthquake – tsunami fast facts. CNN Library. Retrieved from <http://www.cnn.com/2013/07/17/world/asia/japan-earthquake---tsunami-fast-facts/index.html>.

Das, P., Tamminga, K., 2012. The Ganges and the GAP: an assessment of efforts to clean a sacred river. Sustainability 4, 1647–1668. doi: 10.3390/2u4081647. Retrieved from <http://www.mdpi.com/2071-1050/4/8/1647>.

DeLong, J.B., 1998. Estimating world GDP, one million B.C. – present. Department of Economics, U.C. Berkeley. Retrieved from <http://holtz.org/Library/Social%20Science/Economics/Estimating%20World%20GDP%20by%20DeLong/Estimating%20World%20GDP.htm>.

EIA, 2013. International Energy Outlook 2013. U.S. Energy Information Administration. Retrieved from <http://www.eia.gov/forecasts/ieo/world.cfm>.

FAO, 2006. Livestock's long shadow. Food and Agriculture Organization of the United Nations. Retrieved from <http://www.fao.org/docrep/010/a0701.htm>.

Gardner, G., Stern, P., 1996. Environmental Problems and Human Behavior. Allyn and Bacon, Boston.

Harvey, F., Macalister, T., 2014, January 15. BP study predicts greenhouse emissions will rise by almost a third in 20 years. The Guardian. Retrieved from <http://www.theguardian.com/business/2014/jan/15/bp-predicts-greenhouse-emissions-rise-third>.

Healy, P., 2013, July 6. Vast reservoirs of Arctic carbon could affect global warming. NBCLA. Retrieved from <http://www.nbclosangeles.com/news/local/Vast-Reservoirs-of-Arctic-Carbon-Could-Affect-Global-Warming-213926781>.

Heinberg, R., 2011. The End of Growth: Adapting to Our New Economic Reality. New Society Publishers, Gabriola Island, BC.

Herzog, T., 2009, July. World greenhouse gas emissions in 2005. World Resources Institute. Retrieved from <http://www.wri.org/publication/world-greenhouse-gas-emissions-2005>.

Hoggan, R., 2010. Life expectancy in the paleolithic. Retrieved from <http://paleodiet.com/life-expectancy.htm>.

Hoornweg, D., Bhada-Tata, P., 2012. What a waste: a global review of solid waste management. Urban development series; knowledge papers no. 15. World Bank, Washington. Retrieved from <http://documents.worldbank.org/curated/en/2012/03/16537275/waste-global-review-solid-waste-management>.

IAEA, 2013, November 4. IAEA issues projections for nuclear power from 2020 to 2050. International Atomic Energy Agency. Retrieved from <http://www.iaea.org/newscenter/news/2013/np2020.html>.

International Monetary Fund, 1999, September. World Economic Outlook (WEO) Database. Retrieved from <http://www.imf.org/external/pubs/ft/weo/1999/02/data/>.

International Monetary Fund, 2013, October. World Economic Outlook Database. Retrieved from <http://www.imf.org/external/pubs/ft/weo/2013/02/weodata/weoselagr.aspx>.

IPCC, 2007, November. Climate change 2007 summary for policymakers and working group III: mitigation of climate change. Intergovernmental Panel on Climate Change. Retrieved from <http://www.ipcc.ch/pdf/assessment-report/ar4_syr_spm.pdf> and <http://www.ipcc.ch/publications_and_data/ar4/wg3/en/ch1s1-3.html>.

IPCC, 2013, October. Climate change 2013: the physical science basis. Summary for policymakers. Intergovernmental Panel on Climate Change. Retrieved from <http://www.climatechange2013.org/images/uploads/WGI_AR5_SPM_brochure.pdf>.

Kennedy, D., Gläscher, J., Tyszka, J, Adolphs, R., 2009, August 30. Personal space regulation by the human amygdala. Nature Neuroscience 12, 1226–1227 Retrieved from <http://www.ncbi.nlm.nih.gov/pmc/articles/PMC2753689/pdf/nihms-131528.pdf>.

Life of Earth, 2014, June 18. Agricultural pollution causes. Life of Earth. Retrieved from <http://lifeofearth.org/pollution/agricultural-pollution>.

Maitre, M., Boselli, M., 2009, March 27. FACTBOX-key facts on radioactive waste. Reuters. Retrieved from <http://www.reuters.com/assets/print?aid=USLR93723820090327>.

Malthus, T.R., 1988/1798. Population: The First Essay. University of Michigan Press, Ann Arbor, MI.

McKibben, G., 2010. A new world. In: Heinberg, R., Lerch, D. (Eds.), The Post Carbon Reader: Managing the 21st Century's Sustainability Crises. Watershed Media, Santa Rosa, CA, pp. 43–52.

NASA, 2013, October 30. GISS surface temperature analysis (GISTEMP). National Aeronautics and Space Administration: Goddard Institute for Space Studies. Retrieved from <http://data.giss.nasa.gov/gistemp>.

Neurath, P., 1994. From Malthus to the Club of Rome and Back: Problems of Limits to Growth. Population Control and Migrations. M.D. Sharpe, London.

PricewaterhouseCoopers, 2013, January. World in 2050: the BRICs and beyond: prospects, challenges and opportunities. Retrieved from <http://www.pwc.com/en_GX/gx/world-2050/assets/pwc-world-in-2050-report-january-2013.pdf>.

Quaile, I., 2013, April 2. Polar ice sheets melting faster than ever. DW. Retrieved from <http://www.dw.de/polar-ice-sheets-melting-faster-than-ever/a-16432199>.

RIA Novosti, 2014, April 2. Global warming threatens radioactive waste on Novaya Zemlya. The Arctic, Russian Geographic Society. Retrieved from <http://arctic.ru/news/2014/01/global-warming-threatens-radioactive-aste-novaya-samlya>.

Ryerson, W., 2010. Population: the multiplier of everything else. In: Heinberg, R., Lerch, D. (Eds.), The Post Carbon Reader: Managing the 21st Century's Sustainability Crises. Watershed Media, Healdsburg, CA in collaboration with Post Carbon Institute, Santa Rosa, CA.

Smith, J., Schneider, S., Oppenheimer, M., Yohe, G., Hare, W., Mastrandrea, M., Patwardhan, A., Burton, I., Corfee-Morlot, J., Magadza, C., Fussel, H., Pittock, A., Rahman, A., Suarez, A., van Ypersele, J., 2009, March 17. Assessing dangerous climate change through an update of the Intergovernmental Panel on Climate Change (IPCC): reasons for concern. PNAS 106(11), 4133–4137 Retrieved from <http://www.climate.be/users/vanyp/Article%20PNAS%20 2009%20(Update%20IPCC%20Reasons%20for%20concern)/SmithEtal(2009)(PNAS)NewB urningEmbers(Online0812355106).pdf>.

Sokolov, Y., McDonald, A., 2006. Nuclear power – global status and trends. Nuclear Energy 2006. Retrieved from <http://www.iaea.org/OurWork/ST/NE/Pess/assets/nuclear_energy_Alan_ Sokolov06.pdf>.

The World Bank, 2014. Fertility rate, total (births per woman). Retrieved from <http://data.world-bank.org/indicator/SP.DYN.TFRT.IN>.

United Nations, 1999. The world at six billion. United Nations Department of Economic and Social Affairs, Population Division, New York. Retrieved from <http://www.un.org/esa/population/ publications/sixbillion/sixbillion.htm>.

United Nations, 2013. World population prospects: the 2012 revision. United Nations Department of Economic and Social Affairs, Population Division, New York. Retrieved from <http://data. un.org/>.

US Census Bureau, 2013a. World population: total midyear population for the world: 1950–2050. Retrieved from <http://www.census.gov/population/international/data/worldpop/table_popu-lation.php>.

US Census Bureau, 2013b. International data base: mid-year population by single year age groups. Retrieved from <http://census.gov/population/international/data/idb/region.php>.

US Census Bureau, 2013c. World population: historical estimates of world population. Retrieved from <http://www.census.gov/population/international/data/idb/informationGateway.php>.

Wang, Y. 2013, March 25. More people have cell phones than toilets, U.N. study shows. Time.com. Retrieved from <http://newsfeed.time.com/2013/03/25/more-people-have-cell-phones-than-toilets-u-n-study-shows/>.

WHO, 2014. Air pollution. World Health Organization. Retrieved from <http://www.who.int/ topics/air_pollution/en/>.

World Nuclear Association, 2013, November. Radioactive waste management. Retrieved from <http://www.world-nuclear-org/info/Nuclear-Fuel-Cycle/Nuclear-wastes/Radioactive-Waste-Management/>.

Chapter 7

Applying the Brakes: Factors That Limit Growth

Unrestrained materialism and ecological integrity exist in an absolute contradic-tion. We cannot continue to consume and produce at the velocity we are now. The result is a steady erosion of our well-being and the earth's delicate and complex balance.

– Michael Stone (Stone, 2009)

In 44 BCE when Julius Caesar donned the purple toga of Rome's first perpetual dictator, the 500 year old Roman Republic transformed into its more imposing self: the Roman Empire. After countless campaigns to conquer strange lands, the Empire reached its peak in the second century. It stretched from what was Mesopotamia to the Atlantic coast of Western Europe and from England to the Mediterranean shores of Africa. The darker shaded areas in Fig. 7.1 show its impressive extent. It was an onerous task to connect or to protect these vast expanses, particularly without motorized vehicles, modern weaponry, and elec-tronic communication.

After the second century, the Empire weakened and in 476 its last Emperor was deposed. Researchers postulate numerous reasons for its decline: loss of civic virtue and moral decay; a weakened army; disease; reliance on looting rather than on creating wealth; climate change; lead poisoning; geographic expansion; and a clash between Roman and Germanic people (Tainter, 1988). Fig. 7.2 shows the remains of this once-vibrant civilization and reminds us that collapse is possible.

7.1 ANCIENT CIVILIZATION AND LIMITS TO GROWTH

All these causes no doubt contributed to the fall of the Roman Empire. However, a more encompassing reason illustrates the theme of this chapter: When limits are exceeded, collapse will ensue. Many authors support this theory. In the third century, Cyprian, the Bishop of Carthage who lived at the time the Roman Em-pire was failing, wrote a beautifully descriptive letter about its decay:

"There is a diminution in the winter rains that give nourishment to the seeds in the earth, and in the summer heats that ripen the harvests. ...The mountains,

K.L. Higgins: Economic Growth and Sustainability. http://dx.doi.org/10.1016/B978-0-12-802204-7.00007-4

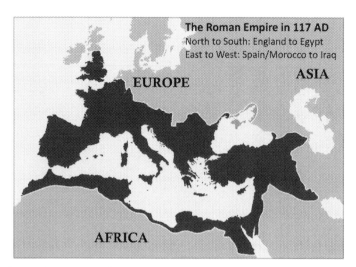

FIGURE 7.1   The Roman Empire in 117 AD. *Source: Adapted from public domain map, released by Ssolbergj, en:User: Andrei nacu. Retrieved from <http://upload.wikimedia.org/wikipedia/commons/b/ba/Roman_Empire_map.svg>.*

disemboweled and worn out, yield a lower output of marble; the mines, exhausted, furnish a smaller stock of the precious metals... Anything that is near its end, and is verging towards its decline and fall is bound to dwindle... This is the sentence that has been passed upon the World...this loss of strength and loss of stature must end, at last, in annihilation."[1]

While they did not personally witness Rome's fall, historians have had time to develop a more philosophical perspective. In the late 1700s, historian Edward Gibbon suggested that the Empire decayed because it grew beyond what its environment could support. Its decline, he says "was the natural and inevitable effect of immoderate greatness. …the causes of destruction multiplied with the extent of conquest;...the stupendous fabric yielded to the pressure of its own weight" (Gibbon, 2005).

In his 1988 work, American anthropologist Joseph Tainter expanded on the resource depletion theory. He argued that "the gradual deterioration or depletion of a resource base ... and the more rapid loss of resources due to an environmental fluctuation or climactic shift ... are thought to cause collapse through depletion of the resources on which a complex society depends." Along with geographic expansion of the Roman Empire, which yielded "a one-time infusion of wealth," came the cost of maintaining its acquisitions for many centuries (Tainter, 1988).

From his considerable research on the subject, social theorist Jeremy Rifkin chimes in on this premise. He suggests that when Rome could no longer

---

1. From Cyprian's *Ad Demetrianum* quoted in Tainter (1988).

The Forum
Rome, Italy

Djémila, Algeria

Hadrian's Wall
Greenhead Lough, England

Bosra, Syria

FIGURE 7.2 Ruins of the ancient Roman Empire. *Source: All photos are public domain licensed under the Creative Commons: The Forum. Retrieved from <http://en.wikipedia.org/wiki/File:Tavares.Forum.Romanum.redux.jpg>; Algeria retrieved from <http://en.wikipedia.org/wiki/File:Djemila_algeria_roman_ruins_145.jpg>; Hadrian's wall retrieved from <http://en.wikipedia.org/wiki/File:Hadrian%27s_wall_at_Greenhead_Lough.jpg>; Syria retrieved from <http://en.wikipedia.org/wiki/File:Roman_ruins,_Bosra,_Syria.jpg>.*

"maintain its empire by new conquests and plunder," it relied on agriculture for 90% of its revenue. Rifkin believes that "the deeper cause of Rome's collapse lies in the declining fertility of its soil and the decrease in agricultural yields" which could no longer sustain its huge infrastructure and its citizens.[2]

These authors, ancient and contemporary, present a systems view of societal decline. Translating their theories into the vernacular of systems thinking, we would say that the Roman Empire exceeded its *limits-to-growth* before it collapsed. Like any natural system, a complex society whose expansion surpasses the ability of its environment to support it, eventually hits a limit. Without adequate resources – food, wealth, or other fuels of growth – such societies decline.

This law of nature still holds true even if we factor in technology. Although the world has dramatically changed in 2000 years, society now faces more

2. Rifkin (2009); statistics on agriculture from Tainter (1988).

severe limits than ever. To understand these limits, we introduce the concept of "carrying capacity."

7.2 CARRYING CAPACITY

Systems expert John Sterman describes the carrying capacity of a system (e.g., Earth) as the condition in which an environment has only so many resources to keep only so many organisms alive. When the number of organisms multiplies or when each one increases its use of resources, the system moves more rapidly toward the upper edges of its *carrying capacity*. In this situation, Sterman notes that the system grows more slowly until scarce resources halt further growth.[3]

Carrying capacity is central to sustainability. It applies to the number of people that Earth can support with its life-sustaining resources. If Earth had inexhaustible carbon-based energy, boundless acres of food-producing land, abundant water, and unlimited ability to absorb pollution, more people would simply imply more towns and businesses; more houses and farms; more water, gas, and oil wells; more consumption of material goods; more employees for growing businesses; and greater well-being for all.

Although our mental model discounts Earth's carrying capacity, we know its capacity is finite. With more demand from a larger population and less supply as we approach resource limits, even low population growth is unsustainable. In other words, Earth's dwindling resources will be inadequate to keep our growing numbers alive and well.

7.2.1 Determinants of Carrying Capacity

Many factors define Earth's carrying capacity. From his extensive study of the subject, Cornell biology professor David Pimentel lists quality and quantity of land, water, energy, and biota (flora and fauna) as resources that determine our ability to survive. He points out that these finite resources must be divided among Earth's people; if population continues to increase, individual shares of these resources will shrink below subsistence levels. His 1999 research estimated that the carrying capacity (Earth's optimum population) is two billion, based on "sustainable use of natural resources" and a European standard of living. This number brings the frightening prospect of drastic adjustment for the seven billion people on Earth today. Pimentel estimated that reducing world population from six billion (the population in 1999) to two billion would require slashing the birth rate to 1.5 children per couple and would take more than a century (Pimentel et al., 1999). Since 1999, estimates will have worsened.

---

3. Sterman (2000). These scarce resources create balancing loops that counteract reinforcing growth loops.

The Global Footprint Network, whose mission is to communicate "the challenges of a resource-constrained world," estimates that in 2007 "humanity used the equivalent of 1.5 Earths to support its consumption" (Ewing et al., 2010). William Ryerson, from the Post Carbon Institute, suggests that if this projection is accurate, Earth is already well over capacity (Ryerson, 2010).

7.2.2   Carrying Capacity and Our Mental Model

Like the strategy that expanded the Roman Empire, our mental model has thrust us closer to Earth's carrying capacity – and perhaps beyond. As individuals who buy things to make ourselves feel good and as nations that rely on energy-dependent economic growth, our complex society operates in an environment of excess. Ignoring Earth's limits is part of our mindset – we do not even contemplate carrying capacity.

Just think about it for a minute. Are you willing to stop buying and using things beyond what you need to survive? And why should you, particularly if you believe that money is your only constraint? Yet, Earth can sustain only so many lives. And remember, because we are dealing with long delay times, growth cannot stop quickly; when it does, it can suddenly reverse into decay.

These gloomy discussions point out the need to acknowledge factors that threaten survival, and to act immediately. Consistent with a holistic view of sustainability, we will pare down our list of limiting factors to four basic resources that maintain our lives and lifestyles: (1) energy supply, (2) water supply, (3) food supply, and (4) forests. Although they may not occupy the same geographic space, these four are ultimately shared by the world.

You may have noticed that although we need air to survive, it is not among these factors. Because limits on air are measured by quality rather than quantity, we included it in our discussion of pollution in Chapter 6. We will return to this element in Chapter 14 when we consider the challenge of managing our shared resources.

7.3   LIMITING FACTORS

Discussions in this section include copious data about the four limiting factors. Again, we do not focus on specific numbers. Instead, we will consider two questions about the trends these data describe:

- Is consumption of these resources increasing?
- Are we approaching their limits?

Although these limiting factors threaten the very notion of sustainability, they are absent from our mental model. Thus, in Chapter 8, we will create a new system diagram that includes these factors so that we can distinguish precursors of collapse and identify possible solutions.

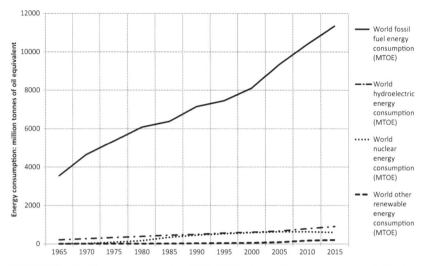

FIGURE 7.3   World energy consumption (1965–2015). *Source: Energy consumption 1965–2010 from BP (2014); projections for 2015 from EIA (2013).*

7.3.1   Energy Supply

For this discussion, we split energy supply into three components: (1) energy from fossil fuels; (2) nuclear energy from enriched uranium; and (3) renewable energy from hydroelectric and other renewable sources. All have limits, although their limits have different characteristics. Fig. 7.3 summarizes consumption for each of these components between 1965 and 2015 and addresses the first question about whether energy consumption is increasing. In a word, the answer is yes. Just in the past decade, aggregate consumption of energy rose 28%.[4]

Our response to the second question regarding constraints is more complex. The following paragraphs describe each energy source and consider this question of limits.

7.3.1.1   Fossil Fuel

It took millions of years for the pressure within Earth's crust to convert carbon-rich remains of plants and animals into fossil fuel. It is taking only a few centuries to burn up these stores of raw energy.

Chapter 6 tells us that our annual consumption of fossil fuels (petroleum, gas, and coal) has more than tripled since 1965. To identify the regions in which the bulk of this increase is occurring, we will use oil consumption as a proxy for fossil fuels; oil accounts for about 38% of fossil fuel consumption and about one-third of the total energy usage (BP, 2014). Fig. 7.4 shows that over the past

---

4. BP (2014); increase occurred between 2003 and 2013.

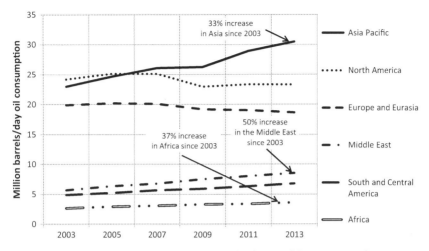

FIGURE 7.4 Regional oil consumption (2003–2013). *Source: Oil consumption data from BP (2014).*

decade, oil consumption declined slightly in North America, Europe, and Eurasia, but rose rapidly in the emerging economies of Africa, Asia Pacific, and the Middle East.

One deviation from these aggregated trends for oil consumption is the consumption of coal. Coal consumption grew an incredible 47% during this same decade; it doubled in the Asia Pacific region and rose 50% in South and Central America (BP, 2014).

Our mental model recognizes that economic growth consumes energy, but omits the fact that fossil fuel is finite. Even so, quantifying our stores of fossil fuels is challenging at best. Although we know about past consumption and can predict future consumption with some confidence, projections for the remaining supplies of fossil fuel or when we will deplete them vary wildly.[5] For instance, although intense media coverage describes the benefits of "fracking" to release natural gas from shale, estimates of its energy contributions are inconsistent. Energy analyst Bill Powers has spent years researching these issues. He calculates that for North America, the likely future production of energy from shale is "about six years of supply at current rates of consumption" (Powers, 2013).

Fig. 7.5 compares oil consumption with oil production. Disparate assumptions about development and discovery create various forecasts for future production (Hagens, 2011). Moreover, energy to develop new sources grows each year as easily accessible supplies dwindle. In other words, it takes much more energy today to generate one unit of energy than it did in the past; expensive equipment and the need to drill deeper to retrieve these buried energy treasures

---

5. Supply includes production capacity, existing energy reserves, and undeveloped resources. For petroleum, a condition called "peak oil" will occur when we reach the maximum rate of extraction.

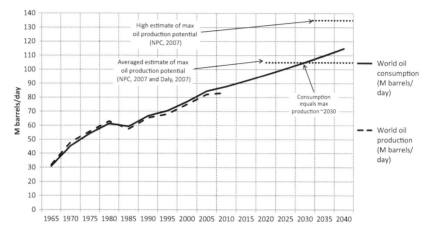

FIGURE 7.5   *Global oil production and consumption (1965–2040). Source: Oil consumption and production data for 1965–2010 from BP (2014); data for 2015–2040 from EIA (2013). Forecasts of max global oil production 2030 from National Petroleum Council (2007) and M. Daly (2007).*

consume more fuel. In the case of US petroleum, energy returned for energy invested (EROEI) has dropped sevenfold in the past 80 years.[6] This change means greater energy consumption, higher production costs, and less to show for our efforts.

7.3.1.2   Nuclear Energy

Although nuclear energy accounted for less than 5% of all energy consumed in 2013,[7] its importance will increase because its immediate effects on the environment are minimal. Chapter 6 reminded us that radioactive waste becomes a danger only when human error or natural disaster compromises nuclear plants or storage facilities.

In Fig. 7.3, nuclear energy consumption has remained relatively flat since the mid-1980s; it decreased nearly 7% between 2011 and 2012, after the earthquake decimated Japan's nuclear plant (see Chapter 6) (BP, 2014). On the supply side, since 1986, nuclear energy capacity has grown far more modestly than fossil fuel production. For instance, between 1990 and 2013, oil production increased 32% (BP, 2014) while nuclear energy capacity increased only 13%.[8] Nuclear capacity may or may not increase by 2030, depending on global policies regarding the safety of nuclear power plants and on competition from other clean renewable energy technologies. Estimates of growth range from 0% to

---

6. Hall and Day (2009); in 1930 the United States used one unit of energy to acquire 100 units; by 2009 one unit returned only 14 units. It is more expensive to extract oil from previously inaccessible areas.

7. BP (2014); nuclear energy consumption was 4.4% of total energy consumption.

8. Capacity of about 330 Gigawatt Electric (GWe) in 1990 estimated from Figure 1 in Sokolov and McDonald (2006); capacity of about 373 GWe in 2013 from IAEA (2013).

94%.[9] If capacity does increase, it will be concentrated in Asia, Eastern Europe, and the Middle East (IAEA, 2013).

A final thought: Nuclear energy relies on enriched uranium. Experts project that uranium supplies should be adequate until 2025, after which new mines will be required. However, plants that convert uranium to uranium hexafluoride for use in nuclear power generation have a more imposing limit; because their capacity is insufficient over the long term, more plants will be required (World Nuclear News, 2013).

7.3.1.3   Renewable Energy

There are various debates about the definition of renewable energy, particularly in the area of biomass. One definition splits renewables into two types: modern renewables and traditional biomass (solid polluting fuel used in rural areas) (REN21, 2014). Because our interest here is in clean energy, we focus on modern renewables and further divide these into two categories: hydroelectric and other renewable energy, which includes geothermal, solar, and wind sources.

Our mental model stresses the importance of clean renewable energy; we count on technology to develop these sources and make their production feasible and efficient. We also rely on nature's gifts of sunlight, wind, precipitation, and other natural bounties as raw materials. These elements, when combined with man-made technologies, define renewable energy's carrying capacity.

Hydroelectricity is the most widely used renewable energy source; in 2010, it provided over 16% of electricity (Worldwatch Institute, 2013; REN21, 2014) and in 2013 it represented 6.6% of all energy consumed.[10] One estimate places future growth of hydroelectric energy at nearly 3% per year into 2040.[11] Other renewable energy (commercially supplied) provided about 2% of total energy consumed and nearly 6% of electricity (BP, 2014; REN21, 2014). Although increased capacity for these renewables is dependent on policy and financing, a significant area of future growth is in household-level energy systems for rural markets that now heat and cook using traditional biomass (REN21, 2014). Fig. 7.3 shows the increase in hydroelectric energy and emphasizes the difference between renewable energy consumption and fossil fuel consumption.[12]

7.3.1.4   Future Limits

Now we can speak to the second question about limits: How close are these energy sources to reaching their limits? The answer is multifaceted; limits for fossil fuels are unlike limits for nuclear, hydroelectric, and other renewable energy.

---

9. World Nuclear News (2013) and NEI (2014). The United States, France, and Russia were the top three nuclear energy generating countries in 2013. As of May 2014, 30 countries had a total of 435 nuclear reactors. In 2013, 72 new nuclear power plants were under construction in 15 countries.

10. BP (2014). Note: various reference materials have different estimates. For example, REN21 (2014) states that hydropower represented 3.8% of final energy consumption in 2012. Data from BP (2014) is used for its historical data and its consistency.

11. EIA (2013); annual growth is projected to be 2.8%.

12. BP (2014); in 2013, other renewable energy sources accounted for 1.9% of total energy.

The fossil fuel limit depends on the size of underground caches. Given the consumption upswing in emerging economies of India, China, and Africa, demand could exceed production by 2030. Extrapolating from Fig. 7.5, a best case scenario pushes this milestone to sometime after 2050. Regardless of exactly when consumption exceeds production, the important point is that in the next 20 to 40 years we may have to rely on reserves rather than on new development. Unfortunately, this possibility prompts more questions: When will new discoveries decline? When must we dig deeper and leave less for future generations?

Limits on nuclear power are more conceptual and financial than physical. Nuclear energy capacity relies on safety, on competition from renewable energy sources, and on expansion of mining operations, uranium processing, and nuclear power plants. These limits can be stretched with technology advances and investment in additional capacity.

For hydroelectric and other renewable energy, the limit is elastic and un-predictable. This energy depends on innovation, on policy support, and on the availability of natural sources such as sunlight, wind, and precipitation. For example, although hydroelectric capacity is projected to increase, we must consider whether climate change-related droughts will impede our ability to generate it.

Currently, other renewable energy is still too embryonic to replace fossil fuels; however, technology will make it more abundant, efficient, and affordable. It is encouraging that the annual global investment in this technology increased nearly sevenfold between 2004 and 2011 when it peaked at $279B. Unfortunately, it fell to $214B in 2013 because of investor concerns about renewable policies and a sharp fall in solar system prices.[13] Even so, renewables have become significant energy sources in some countries. For example, in 2013 wind power provided a third of Denmark's electricity and a fifth of Spain's (REN21, 2014). The discouraging news is that as energy consumption increases and as renewable energy capacity increases, some estimates for 2040 still place commercially supplied renewable energy at the same 1.9% of total use that it is today (EIA, 2013). Thus, holding hope that we can fully transition to renewable energy anytime soon is unrealistic. Its limits may change as technology advances, but they will be far lower than our energy needs.

7.3.2 Water Supply

Our mental model assumes that the most vital fount of physical life on Earth – water – will always be there. In addition to salt-water oceans that make up 97% of Earth's water, two fresh water sources meet our needs. The first, groundwater, comes from underground supplies that feed wells and springs; the second is surface water from precipitation that collects in streams, rivers, and reservoirs.

---

13. Frankfurt School UNEP Centre/BNEF (2014) and REN21 (2014).

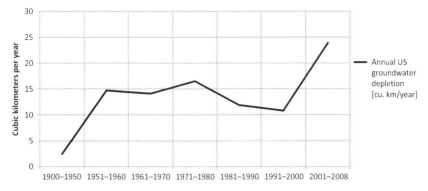

FIGURE 7.6   Annual average groundwater depletion in the United States (1900–2008). *Source: Groundwater depletion data from Konikow (2013).*

7.3.2.1   Groundwater

Just as underground pools of fossil fuels are finite, so are many underground pools of water. In addition to shallow aquifers that are recharged by surface water, the world also relies on fossil water that was trapped deep beneath the ground during the Ice Age. Ancient aquifers under the deserts of Africa and Australia, under the midsection of the United States, and elsewhere are replenished little, if any, by precipitation. These precious stores of groundwater are rapidly shrinking as they nourish our growing populations.

Between 2001 and 2008, more people, more energy production, and more agriculture in the United States soaked up groundwater about three times faster than during the *entire* twentieth century. Total depletion between 1900 and 2008 was about twice the volume of Lake Erie.[14] Fig. 7.6 highlights an abrupt and unprecedented rise in US groundwater depletion in the past decade (i.e., groundwater has decreased).

Those who depend on groundwater for life and livelihood are worried. For example, Kansas farmers who use water from the Ogallala Aquifer find that their wells are producing half as much as they did 10 years ago; they are investigating conservation techniques (Farabaugh, 2014).

7.3.2.2   Surface Water

Surface water irrigates many of our farms, supports communities and businesses, and feeds hydroelectric dams. Lack of surface water will therefore diminish food supplies and affect the availability and cost of electricity.

In recent years, surface supplies of fresh water have been strained by severe weather. Rivers, lakes, and reservoirs are shrinking in many areas including the

---

14. Konikow (2013). U.S. Geological Survey conducted this research. Phone interview with Konikow from Zabarenko (2013). From 2001 to 2008, groundwater depletion rate averaged $\sim$25 km^3 per year versus 9.2 km^3 per year from 1900 to 2008. Depletion of US groundwater 1900–2000 was about 800 cubic kilometers; by 2008 it was 1000 cubic kilometers, a 25% increase in just 8 years.

mid and western United States, Mexico, Argentina, Spain, Southeastern China, Iraq, and Kenya where droughts have reached near record levels (Tirado and Cotter, 2010). Surface water shortages, in turn, increase dependency on groundwater and further exhaust its supplies.

In 2012, the devastating drought in southwest China reached its third year and had residents hiking 6 miles to transport drinking water. Damaged crops and energy shortages boosted their living expenses (Zhang, 2012). In January 2014, California drought reached extreme levels. Not only did it affect drinking water, intensify fire danger, and increase local unemployment, it also reduced food supplies for other US states and other countries. Ranchers there are losing their livelihoods; many had to sell their thirsting cattle at cut rates.[15] And water shortages in these areas have already prompted water theft (Cervantes, 2014). If global warming does create weather extremes, droughts like these will no longer be rare.

7.3.2.3 Future Limits

Since the 1960s, the world has used more underground and surface water than is replenished by nature. Because annual groundwater depletion rates have doubled,[16] limits to water supplies are already affecting us. By 2050, researchers expect that over 50%, or about five billion, of the world's people will live in water stressed areas, particularly in India, Northern Africa, and the Middle East (Roberts, 2014).

There is another side to the decline in regional water supplies. As various regions become stressed, water will become more precious everywhere. Now enter the concept of what economists call "virtual water" – the water required to grow food. Agriculture (mostly food) drinks up an estimated two-thirds of the world's water. About a tenth of these products are exported (Pearce, 2008). Thus, as more regions experience water shortages, they will not only seek water; they will also import food since their parched lands cannot support water-thirsty agriculture.

When countries that now export virtual water (food) reconsider this strategy, food and water supplies will decline for some and drive up prices for all. Food and water shortages also create tension and conflict as nations compete for these life-giving assets (Pearce, 2008). To put things into perspective, it takes about 1000 tons of water to grow a ton of wheat, 65 gallons per pound of potatoes, and 650 gallons per pound of rice.

Now, as a final thought, let us consider the competition between two primary resources: groundwater and fossil fuel. As we approach the limits of existing sources and search for new ones, a major battle is brewing in the United States.

---

15. Park and Lurie (2014); Pierson (2014). California grows about half the fruits, nuts, and vegetables and is the leading wine and dairy producer in the United States.
16. Wada et al. (2010). Annual rate of groundwater abstraction climbed from 312 cu. km/year in 1960 to about 734 cu. km/year in 2000; groundwater depletion grew from 126 cu. km/year to about 283 cu. km/year over the same period.

The company TransCanada and several US states want to build an oil pipeline from Canada to Texas. This endeavor promises to create jobs, increase profits, and distribute oil supplies. Farmers and other environmentalists fear that in a few years, the oil and added chemicals in the pipeline will leak into the precious Ogallala aquifer and poison agricultural and drinking water for two million people in eight states. The issue is so contentious that it has been taken to court (Reynolds, 2014). We should expect worldwide conflicts of this nature in the future.

7.3.3 Food Supply

Although numbers vary slightly with volcanoes, rising sea levels, and land reclamation, about 29% of the Earth is land and 71% is water (CIA, 2001). Food-producing land comprises about 38% of Earth's land.[17] We share these statistics only to emphasize that the physical limit for food-producing resources is real; it is tied to the size of our planet.

Two sources of food supply are land that supports crops and livestock, and water that nurtures fisheries. Population growth and pollution threaten both. In the quest to make these food sources more productive and accommodate an ever-increasing number of people, urbanization has encroached on productive land; pollution has degraded soil and has contaminated cropland and fish habitats. Nevertheless, increased productivity from better crop rotation, fertilizers, and pesticides has produced more food. More fish have been harvested as the number and size of fishing vessels increase.

7.3.3.1 Food-Producing Land and Fisheries

Three types of land produce food. Arable land is cultivated for crops; pastures support livestock; and permanent croplands grow products such as coffee, rubber, fruit, cocao, and nuts. Except for an increase in the proportionally tiny permanent cropland, Fig. 7.7 shows that the world's productive land declined slightly between 2000 and 2008 while world population increased 10%.

Fisheries are a different story. Although reliable data for the number or volume of fisheries is elusive, we do know about fish populations, or "stocks" of various types of fish. As we will soon see, fish production from these stocks has risen.

Even though arable land has decreased slightly and fisheries have been polluted, the number of undernourished people in the world has fallen by 17% since 1990–1992 (FAO, 2013). At first blush, this all sounds like good news – nearly the same land, the same water, and more food. However, recall from Chapter 6 that accumulated damage from pollution is making some land and water unproductive. Furthermore, two additional concerns are hidden by these aggregates. First, fish production is moving from stocks that are overfished (the number of

---

17. FAOSTAT (2010). Earth's total land area in 2008 was about 13 billion hectares; arable land = 1.38 billion, permanent cropland = 0.15 billion, and pastures = 3.36 billion hectares. One hectare = 0.01 sq. km or 2.47 acres.

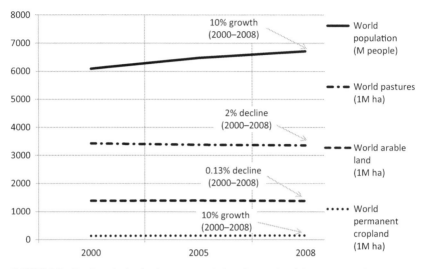

FIGURE 7.7   Food-producing land versus population. *Source: Land data from FAOSTAT (2013). Population data from US Census Bureau (2013).*

fish is decreasing and cannot be sustained) to those that are underfished (abundant fish can sustain their numbers), that is, we are catching different types of fish. Second, deforestation adds to food-producing land as pollution diminishes it. We discuss these concerns later in the chapter.

7.3.3.2   Changes in Food Production

Fig. 7.8 shows production of selected food types between 1965 and 2010: (1) fish, grain, corn, and rice production has risen faster than population while arable land has remained nearly constant since about 1985; (2) cattle and buffalo production has modestly increased as more people eat more meat; and (3) fish production has grown substantially.

There is another concern embedded in these statistics. While fish production more than tripled since 1965, the source of fish has changed dramatically. Its growth comes from "aquaculture" (man-made fisheries in ponds, lakes, rivers, and oceans); production from capture fisheries (fish that grow in the wild) has barely changed. Between 2001 and 2011, capture fish inched up from 92 to 93 million tons while aquaculture fish production increased more than 60% from 38 to 63 million tons (FAO, 2014a).

7.3.3.3   Future Limits

Of course, food supply affects population growth, specifically through hunger and starvation. Thankfully, with the hundreds of international efforts to improve the state of food insecurity (e.g., the United Nations' organizations and non-government organizations such as "Action Against Hunger"), the number of

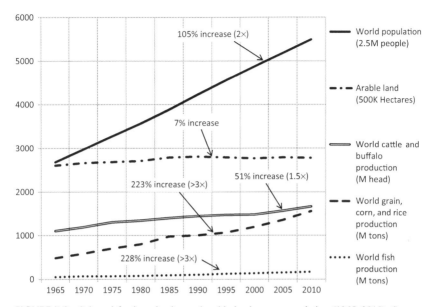

FIGURE 7.8   Selected food production and arable land versus population (1965–2010). *Source: Grain, corn, rice production, cattle and buffalo production, and arable land from FAOSTAT (2013); fish production from FIGIS (2014); world population from US Census Bureau (2013).*

undernourished people in the world continues to drop. However, "the rate of progress appears insufficient to reach international goals for hunger reduction in developing regions." People who suffer most live in Southern and Eastern Asia and sub-Saharan and Eastern Africa (FAO, 2013). Unless food production continues to rise as rapidly as it has in the past 50 years, starvation will eventually halt population growth. Now we ask what might prevent sustained increase in food production. The following examples illustrate two of the many roadblocks.

First, there is grave concern that we are destroying our limited arable land. Agricultural waste coupled with poor irrigation design has leached salt from irrigated fields and brought it to the surface. Salinization and waterlogging have degraded an estimated 10–15% of the 260 million hectares of irrigated soil; about 1.5 million hectares disappear from productive use each year. In Egypt, India, and Syria, degradation is even higher – between 30% and 50%. Some cotton fields in Central Asia have reached catastrophic levels (Carley and Christie, 2000).

A second concern relates to fish supplies. Earlier we noted that fish production is migrating; between 1974 and 2011, underfished stocks dropped from 40% to 15% of total fish stocks while overfished stocks increased from 10% to 29%. Thus, the apparent increase in fish production is only the result of "fisheries moving to underfished resources as others become overfished and depleted" (FAO, 2014b).

Although food production from land and water appears to be keeping pace with population, we are dipping into our reserves; we are "robbing Peter to pay Paul." This strategy is unsustainable. In other words, we cannot sit back and expect that current growth trends for either food production or population will continue indefinitely.

7.3.4 Forests

Forests make up about 31% of our landmass (FAO, 2012). They are integral to our ecosystem and woven into our lifestyles. The list of tree-related products and benefits – from building materials to the pleasure of being with nature – is long. Forests are fourth on our list of limited resources.

7.3.4.1 Deforestation

In 2010, the total forest area in the world was just above four billion hectares, with over half of that in the Russian Federation, Brazil, Canada, the United States, and China. Each year, forest area decreases for reasons that include fires, drought, and pest infestation. The greatest cause, however, is conversion of tropical forests to agricultural land.[18] For example, pastures now occupy some 70% of deforested land in the Amazon and feed crops are grown on much of the remaining 30% (FAO, 2006).

Ecologically speaking, by regulating CO_2, forests maintain the delicate balance of the atmosphere. Living forests absorb CO_2 into their wood, leaves and soil; 2009 estimates suggest that global forests absorbed over a hundred times the CO_2 emissions of the United States (Congressional Budget Office, 2012). However, as forests vanish, CO_2 levels rise and exacerbate global warming. Global warming increases potential for fires and drought, which can destroy more forestland. Furthermore, when forests are cleared, their stored carbon escapes into the atmosphere. All in all, deforestation accounts for 12% of total CO_2 emissions (Herzog, 2009). Accordingly, as population growth and economic growth generate more CO_2 and diminish forests, nature eliminates less.

Between 1990 and 2010, the world lost over 135 million hectares (334 million acres) of forests (see Fig. 7.9). In 2010, the estimated annual rate of deforestation was 5.2 million hectares (about the size of Costa Rica) (FAO, 2012). Tropical rainforests, such as those in Brazil and Indonesia, are disappearing at a rate of 80,000 acres per day (Butler, 2012). Although deforestation rose in countries such as Australia, overall the rate of deforestation declined in the 2000s compared with the 1990s – partly a result of Brazil's focus on conservation and reforesting (FAO, 2012).

Diminishing Earth's ability to clean the atmosphere is serious enough, but as tropical forests over twice the size of Ireland are razed each year, Earth faces

---

18. Data from FAO (2012).

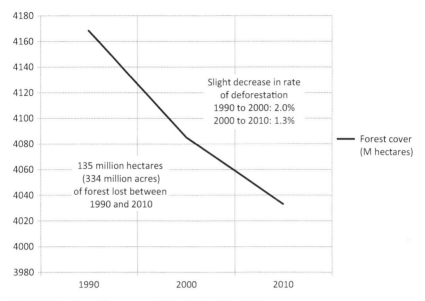

FIGURE 7.9   Global forest cover (1990–2010) (FAO, 2012).

extinction of a million plant and animal species (Gardner and Stern, 1996). In fact, about 50% of the world's terrestrial species live in the rainforests (Butler, 2012). Concerns about reduced biodiversity and its effects on ecological balance are another topic altogether – critically important, but beyond the scope of this book.

All these statistics are startling to say the least. So now, what about the future of forests and their relationship to sustainability?

7.3.4.2 Future Limits

By itself, deforestation merely reflects a growing population and its mounting need for wood products and food-producing land. If fossil fuels were limitless and their by-products were harmless, or if there were other ways to regulate CO_2 and we did not care about having forests in the world, these trends would be of little concern. However, these "ifs" are untrue. Increased consumption of carbon-based energy emits more greenhouse gas at the same time that Earth's cleansing forests are shrinking; both phenomena contribute to weather extremes.

The good news is that we have seen remarkable progress in preserving forests. Conservation and large-scale planting of trees in some countries have reduced the annual loss of forests by over a third.[19] Attention to our forests will be an important part of our strategy to achieve sustainability. As we will see in

19. FAO (2012). Annual deforestation rate decreased from 8.3 million hectares (1990–2000) to about 5.2 million (2000–2010).

Chapter 12, many countries have already pledged to reduce greenhouse gas by reducing deforestation.

This chapter confronts the reality of limits. Limits put on the brakes in a system. They are a critical aspect of sustainability. In systems thinking terms, they are part of the balancing loops that rein in the growth or decay generated by reinforcing loops.

We have so far identified four factors that can limit Earth's carrying capacity, that is, its ability to sustain our increasing population. In describing these factors, we answered two questions about limits. First, for the most part, consumption of energy, water, food, and forests is increasing. For some resources, limits are hard and physical (e.g., fossil fuel and groundwater); for others, limits depend on technology and policies (e.g., nuclear energy) or on our relationship with Earth's ecosystem (e.g., surface water and arable land). We have seen that as population increases, depletion of these resources rises. Yet, we mask their limits by dipping into reserves (fossil fuels and groundwater), by depleting one source then migrating to another (fisheries), by polluting land and water to increase productivity (fertilizing arable land), or by destroying one resource to create another (deforestation). These strategies threaten future life.

If we continue speeding toward these limits with our current addiction to growth, population will eventually plummet as water, food, and energy shortages skyrocket. Economies will plunge when robbed of the energy resources that keep them going. We can take heed from the fall of the Roman Empire and its inattention to an environment that could no longer support its complexity. Achieving sustainability in the twenty-first century depends on how quickly we manage Earth's carrying capacity and how well we compensate for the harmful conditions we have already created.

It does not much matter whether one's home nation or state or community has enough water or food, or enough energy, or enough forests. We will *all* be affected in some way. We all live on the same Earth whose ability to support us is finite.

REFERENCES

BP, 2014, June. BP statistical review of world energy 2014. Retrieved from <http://www.bp.com/en/global/corporate/about-bp/energy-economics/statistical-review-of-world-energy.html>.

Butler, R., 2012, July 17. Deforestation in the Amazon. Mongabay.com. Retrieved from <http://www.mongabay.com/brazil.html>.

Carley, M., Christie, I., 2000. Managing Sustainable Development, second ed. Earthscan, London.

Cervantes, E., 2014, June 20. Valley drought 2014: thieves steal 2,500 gallons of water. KMPH Fox 26. Retrieved from <http://www.kmph-kfre.com/story/25834290/valley-drought-2014-thieves-steal-2500-gallons-of-water>.

CIA, 2001, July. Long-term global demographic trends: reshaping the geopolitical landscape. Central Intelligence Agency. Retrieved from <https://www.cia.gov/library/reports/general-reports-1/Demo_Trends_For_Web.pdf>.

Congressional Budget Office, 2012, January. Deforestation and greenhouse gases. The Congress of the United States. Retrieved from <http://www.cbo.gov/sites/default/files/cbofiles/attachments/1-6-12-forest.pdf>.

Daly, H., 2007. Ecological Economics and Sustainable Development, Selected Essays of Herman Daly. Edward Elgar, Northampton, MA.

EIA, 2013. International Energy Outlook 2013. U.S. Energy Information Administration. Retrieved from <http://www.eia.gov/forecasts/ieo/world.cfm>.

Ewing, B., Moore, D, Goldfinger, S., Oursler, A, Reed, A, Wackernagel, M., 2010, October 13. The Ecological Footprint Atlas 2010. Global Footprint Network, Oakland. Retrieved from <http://www.footprintnetwork.org/images/uploads/Ecological_Footprint_Atlas_2010.pdf>.

FAO, 2006. Livestock's long shadow. Food and Agriculture Organization of the United Nations. Retrieved from <http://www.fao.org/docrep/010/a0701.htm>.

FAO, 2012, December 17. Global forest resources assessment 2010. Food and Agriculture Organization of the United Nations. Retrieved from <http://www.fao.org/forestry/fra/fra2010/en/>.

FAO, 2013. The state of food insecurity in the world: the multiple dimensions of food security. Executive Summary. Food and Agriculture Organization of the United Nations. Retrieved from <http://www.fao.org/docrep/018/i3458e/i3458e.pdf>.

FAO, 2014a. Summary tables of fishery statistics. Food and Agriculture Organization of the United Nations. Retrieved from <ftp://ftp.fao.org/FI/STAT/SUMM_TAB.HTM>.

FAO, 2014b. The state of world fisheries and aquaculture. Food and Agriculture Organization of the United Nations. Retrieved from <http://www.fao.org/3/a-i3720e.pdf>.

FAOSTAT, 2010. FAO Statistical Yearbook 2010. Table A4: land use. Retrieved from <http://www.fao.org/economic/ess/ess-publications/ess-yearbook/ess-yearbook2010/en/>.

FAOSTAT, 2013. FAO Statistical Yearbook 2013. Food and Agriculture Organization of the United Nations. Retrieved from <http://faostat.fao.org/>.

Farabaugh, K., 2014, March 21. Kansas farmers work to prevent depletion of Ogallala Aquifer. Voice of America. Retrieved from <http://www.voanews.com/content/kansas-farmers-work-to-prevent-depletion-of-ogallala-aquifer/1876717.html>.

FIGIS, 2014. Global production statistics 1950–2012 (fish). Food and Agriculture Organization of the United Nations. Retrieved from <http://www.fao.org/fishery/statistics/global-production/query/en>.

Frankfurt School UNEP Centre/BNEF, 2014. Global trends in renewable energy investment 2014, key findings. Frankfurt School of Finance & Management, Frankfurt, GE. Retrieved from <http://fs-unep-centre.org/sites/default/files/attachments/14008nef_visual_12_key_findings.pdf>.

Gardner, G., Stern, P., 1996. Environmental Problems and Human Behavior. Allyn and Bacon, Boston.

Gibbon, E., 2005. In: Womersley, D. (Ed.), The History of the Decline and Fall of the Roman Empire, abridged ed. Penguin Group, London.

Hagens, N., 2011, August 24. Low carbon and economic growth: are both compatible in developing economies? Post Carbon Institute. Retrieved from <http://www.postcarbon.org/blog-post/463543-low-carbon-and-economic-growth-are>.

Hall, C., Day, Jr. J., 2009. Revisiting the limits to growth after peak oil. American Scientist 97, 230–237. Data courtesy of the Association for the Study of Oil and Gas. Retrieved from <www.esf.edu/efb/hall/2009-05Hall0327.pdf>.

Herzog, T., 2009, July. World greenhouse gas emissions in 2005. World Resources Institute. Retrieved from <http://www.wri.org/publication/world-greenhouse-gas-emissions-2005>.

IAEA, 2013, November 4. IAEA issues projections for nuclear power from 2020 to 2050. International Atomic Energy Agency. Retrieved from <http://www.iaea.org/newscenter/news/2013/np2020.html>.

Konikow, L., 2013. Groundwater depletion in the United States (1900–2008). Scientific Investigations Report 2013–5079. U.S. Geological Survey. Retrieved from <http://pubs.usgs.gov/sir/2013/5079/SIR2013-5079.pdf>.

National Petroleum Council, 2007. Hard truths: facing the hard truths about energy: a comprehensive view to 2030 of global oil and natural gas. Retrieved from <http://npchardtruthsreport.org/download.php>.

NEI, 2014. World statistics: nuclear energy around the world. Nuclear Energy Institute. Retrieved from <http://www.nei.org/Knowledge-Center/Nuclear-Statistics/World-Statistics>.

Park, A., Lurie, J., 2014, February 10. California's drought could be the worst in 500 years. Mother Jones. Retrieved from <http://www.motherjones.com/environment/2014/02/california-drought-matters-more-just-california>.

Pearce, F., 2008, June 19. Virtual water. Forbes. Retrieved from <http://www.forbes.com/2008/06/19/water-food-trade-tech-water08-cx_fp_0619virtual.html>.

Pierson, D., 2014, February 4. Drought leaves dark cloud over California ranchers, growers. Los Angeles Times. Retrieved from <http://articles.latimes.com/2014/feb/04/business/la-fi-ranchers-drought-20140205>.

Pimentel, D., Bailey, O., Kim, P., Mullaney, E., Calabrese, J., Walman, L., Nelson, F., Yao, X., 1999. Will limits of the earth's resources control human numbers? Environment, Development and Sustainability 1, 19–39.

Powers, B., 2013. Cold. Hungry and in the Dark: Exploding the Natural Gas Supply Myth. New Society Publishers, Gabriola Island, BC.

REN21, 2014. Renewables 2014 global status report. Renewable Energy Policy Network for the 21st Century. Retrieved from <http://www.ren21.net/Portals/0/documents/Resources/GSR/2014/GSR2014_full%20report_low%20res.pdf>.

Reynolds, D., 2014, May 20. Nebraska becomes battlefield in fight over Keystone XL pipeline. CBS News. Retrieved from <http://www.cbsnews.com/news/nebraska-becomes-battlefield-in-fight-over-keystone-pipeline/>.

Rifkin, J., 2009. The Empathic Civilization: The Race to Global Consciousness in a World in Crisis. Penguin Group, New York.

Roberts, A., 2014, January 9. Predicting the future of global water stress. MIT News. Retrieved from <http://web.mit.edu/newsoffice/2014/predicting-the-future-of-global-water-stress.html>.

Ryerson, W., 2010. Population: the multiplier of everything else. In: Heinberg, R., Lerch, D. (Eds.), The Post Carbon Reader: Managing the 21st Century's Sustainability Crises. Watershed Media, Healdsburg, CA in collaboration with Post Carbon Institute, Santa Rosa, CA.

Sokolov, Y., McDonald, A., 2006. Nuclear power – global status and trends. Nuclear Energy 2006. Retrieved from <http://www.iaea.org/OurWork/ST/NE/Pess/assets/nuclear_energy_Alan_Sokolov06.pdf>.

Sterman, J., 2000. Business Dynamics: Systems Thinking and Modeling for a Complex World. Irwin McGraw-Hill, New York.

Stone, M., 2009. Yoga for a World out of Balance: Teachings on Ethics and Social Actions. Shambhala, Boston.

Tainter, J., 1988. The Collapse of Complex Societies. Cambridge University Press, Cambridge.

Tirado, R., Cotter, J., 2010, April. Ecological farming: drought-resistant agriculture, Greenpeace Research Laboratories, University of Exeter, UK. GRL-TN 02/2010. Retrieved from <http://www.greenpeace.org/international/Global/international/publications/agriculture/2010/Drought_Resistant_Agriculture.pdf>.

US Census Bureau, 2013. World population: total midyear population for the world: 1950–2050. Retrieved from <http://www.census.gov/population/international/data/worldpop/table_population.php>.

Wada, Y., van Beek, L., van Kempen, C., Reckman, J., Vasak, S., Bierkens, M., 2010. Global depletion of groundwater resources. Geophysical Research Letters 37 (L20402, 3) 1–5. Retrieved from <http://tenaya.ucsd.edu/~tdas/data/review_iitkgp/2010GL044571.pdf>.

World Nuclear News, 2013, September 12. Uranium supply and demand in balance for now. Retrieved from <http://www.world-nuclear-news.org/ENF-Uranium_supply_and_demand_in_balance_for_now-1209137s.html>.

Worldwatch Institute, 2013. Use and capacity of global hydropower increases. Retrieved from <http://www.worldwatch.org/node/9527>.

Zabarenko, D., 2013, May 20. Drop in U.S. underground water levels has accelerated: USGS. Reuters. Retrieved from <http://www.reuters.com/article/2013/05/20/us-usa-water-idUSBRE-94J0Y920130520>.

Zhang, Y., 2012, April 5. China drought 2012: three-year-long dry spell continues in Southwest. International Business Times. Retrieved from <http://www.ibtimes.com/china-drought-2012-three-year-long-dry-spell-continues-southwest-554974>.

Chapter 8

The "I"s Have It: A Systems View of Sustainability

...linkage of differentiated parts of a system – is at the heart of well-being.

– Dan Siegel (Siegel, 2012)

We are trying to undo some of the harm we have done, and as climate change worsens we will try harder, even desperately, but until we see that the Earth is more than a mere ball of rock we are unlikely to succeed.

– James Lovelock (Lovelock, 2009)

"Integration" and "interdependence" – two small words with big implications. These tiny treasures hold the secret to sustainability. By viewing Earth as a single entity – an intertwined system of countless components – we can appreciate the significance of these "i" words. We are all part of this same system, this same planet; our survival relies on all parts working well and working together.

8.1 INTEGRATION AND INTERDEPENDENCE

If you are one of the tens of millions of people who saw James Cameron's 2009 film *Avatar,* you have experienced a powerful image of the first "i" word: "integration." This futuristic movie astonished audiences with its special effects and message of connection. A massive biological neural network – the giant willow-like Tree of Souls – united the thoughts and memories of all Na'vi people on the planet Pandora as it linked them with their spiritual roots. Their minds were woven into a single consciousness.

The second "i" word, "interdependence," has long been implied in human notions about the world. Ancient civilizations intuitively knew that their lives were entwined with Earth's bounties; the Greeks believed Gaia was their benefactress (Fig. 8.1), the Slavs attributed their welfare to Mat Zemlya, and others relied on good old Mother Earth. Although these early beliefs lacked scientific sophistication, they did reveal profound awareness of our interdependence with Earth.

Since the 1920s when Russian biogeologist Vladimir Vernadsky observed that life was connected to Earth's self-contained biosphere, other science-based theories have expressed similar relationships (Vernadsky, 1998). One

K.L. Higgins: Economic Growth and Sustainability. http://dx.doi.org/10.1016/B978-0-12-802204-7.00008-6

95

FIGURE 8.1 Gaia goddess of Earth. *Source: Detail of Gaia from a painting signed by Aristophanes, ca. 401–400 BCE; drawn by Wilhelm Heinrich Roscher Creative Commons. Retrieved from <http://upload.wikimedia.org/wikipedia/commons/c/c2/Gaia_Aristophanes.JPG>.*

well-known example, the Gaia hypothesis (named after the Greek goddess), was introduced in the 1970s by British environmentalist James Lovelock and American biologist Lynn Margulis. Although parts of this hypothesis have been controversial, its central theme of interdependence is highly relevant today. It describes Earth as "a single physiological system, an entity that is alive" and self-regulates at a favorable state for its inhabitants (Lovelock, 2000). In other words, life interacts with an earthly system whose parts collaborate to sustain that life.

8.2 THE SYSTEM DIAGRAM

This chapter draws from the message of *integration* from *Avatar* and the theme of *interdependence* from the Gaia hypothesis to paint the sustainability picture anew. It combines trends from previous chapters into a new system diagram – an interlocking portrait of environment, economy, and society. The chapter also reveals imperfections of this interpretation and prepares us for the leap from thoughts and concerns to actions.

For ease of understanding, the new system diagram comes together in two steps. The first step integrates economy with environment. The second step brings in society, incorporates sustainable happiness and well-being from Chapter 5, and emphasizes the interaction among all parts of the sustainability triad.

The many loops and arrows in the system diagrams look daunting at first, but we will describe and summarize each relationship as we go. Remember that reinforcing loops represent growth or decay. They are like a car that picks up speed as it coasts downhill. Balancing loops are the steering wheels that hold the system to a particular goal or limit. Carrying capacity represents the limits of the entire system; a combination of many balancing loops may be involved in this concept. When a system reaches its carrying capacity, growth or decay may slow, stop, or even reverse.

Recall from Chapter 1 that an "s" (same) at the head of an arrow indicates that the two elements (cause and consequence) move in the same direction; an "o" (opposite) means that when cause goes up, consequence goes down, and vice versa. To prevent confusion, the descriptions italicize the names of each element and each loop as they appear in the diagrams. Finally, note that the dashed boundary line that separated internal relationships from external influences in the mental model is absent. The system now includes the entire Earth and, for our purposes, has no relevant externalities.

Once you appreciate the individual interactions described in the following sections, hold the final diagram at arm's length and note the spaghetti tangle of interdependencies. You may wish to pick an area of interest and ask: What will happen in the rest of the system if this element increases or decreases? Just a few questions will reveal the complexity of sustainability and the dysfunctional path that unrestrained economic and population growth are carving for us.

8.2.1   Step 1: Integrating Economy and Environment

The system diagram in Fig. 8.2 expands on our mental model. It illustrates how economic growth harms the environment and how environmental damage

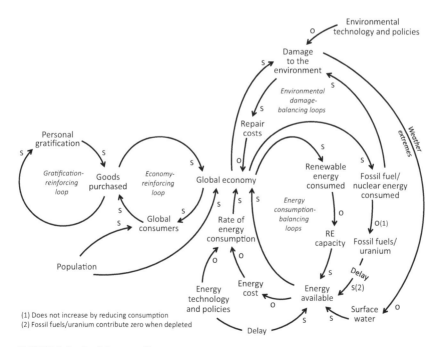

FIGURE 8.2   Partial system diagram (economy and environment).

diminishes the economy. Finally, it incorporates limits to the energy resources that fuel the economy.

8.2.1.1 Economy and Environmental Damage

Our discussion begins in the center of Fig. 8.2 with the *global economy* and the *economy-reinforcing loop* from our mental model; it shows the same positive relationship between *population* and the number of *global consumers* and between *population* and the *global economy*. Growth of the *global economy* still enables more *global consumers* to purchase more *goods*; increased *population* improves the *global economy* when more people constructively contribute and when nations and organizations stimulate innovation and increase employment opportunities. Different from the mental model, growth of the *global economy* causes *damage to the environment*. It generates pollution including greenhouse gas (GHG) that permeates the atmosphere and contributes to *weather extremes,* and other wastes that seep into water supplies, poison the air, or end up in the ground. The *gratification-reinforcing loop* on the left operates in the short term as it did in the mental model: More *goods purchased* increases *personal gratification* which, when buyers find themselves unsatisfied, stimulates more *goods purchased.*

Summary:

- The economy is stimulated in part by personal gratification and people.
- Economic growth that depends on fossil fuels harms the environment.

8.2.1.2 Energy Consumption: Balancing Loops and Limits

As in our mental model, growth of the *global economy* increases *energy consumed*; economic decay decreases *energy consumed*. This diagram replaces the single *energy consumption-reinforcing loop* with two *energy consumption-balancing loops*: one for *renewable energy* and the other for nonrenewable *fossil fuel/nuclear energy*. These loops define one aspect of Earth's carrying capacity, that is, the limit of its ability to support our lives.

Although these two *energy consumption-balancing loops* have always been present, our mental model ignored them – and for good reason. They are currently too weak to overpower the economic growth machine. The possibility of depleting Earth's fossil fuels seems too remote to capture our attention.

Following the flow of these balancing loops, a rise in the *global economy* increases energy consumption (*renewable energy consumed* and *fossil fuel/nuclear energy consumed*). In the renewable energy loop, more *renewable energy consumed* reduces *RE capacity* (the current state of renewable energy capacity), which in turn decreases *energy available* (until technology adds renewable energy capacity). Less *energy available* diminishes the *global economy* once known efficiency measures have been applied and businesses are forced to cut back operations.

The balancing loop for *fossil fuel/nuclear energy* operates similarly with two exceptions. First, increased *fossil fuel/nuclear energy consumption* produces pollution such as GHG, air pollution, and radioactive waste, which cause *damage to the environment*. And second, increased consumption depletes *fossil fuels/uranium*. When these fuel sources are gone, their contribution to *energy available* becomes zero and we must then depend entirely on *renewable energy* to fuel economic growth.

The final relationship in this group shows that reducing *energy available* increases *energy cost* (supply and demand). Then, theoretically, when *energy cost* rises, conservation decreases the *rate of energy consumption*, which dampens the *global economy*.

Summary:

- Economic growth requires energy, most of which today comes from finite sources.
- When fossil fuels and uranium are depleted, the economy must depend on renewable energy whether or not there is adequate capacity.
- As energy sources are depleted, energy costs rise and spur conservation.

8.2.1.3   Consequences of Environmental Damage on the Economy

Damage to the environment affects the economy in two ways. First, as part of an *environmental damage-balancing loop*, it siphons GDP away from productive investments such as education or innovation and uses it to pay *repair costs* of cleaning up waste. Over the long term, these costs depress the *global economy*, which reduces both *damage to the environment* and *fossil fuel/nuclear energy consumed*.

Second, *damage to the environment* in the form of excess GHG contributes to global warming, which produces *weather extremes*. *Weather extremes* intensify damage from hurricanes and coastal storms, and elevate repair costs. In addition, *weather extremes* that cause drought diminish *surface water* used to generate hydroelectric power, thus decreasing *energy available*; those that cause bitter cold or searing heat increase the energy needed to stay warm or cool, which also reduces *energy available* (direct link not shown). By having less water-powered renewable *energy available*, we rely more on *fossil fuels/ uranium* and must add energy generation plants to meet increased needs. We then more quickly approach the *fossil fuels/uranium* limits of the system that, over time, depresses *economic growth*.

Summary:

- Environmental damage causes weather extremes, reduces energy available, and depletes limited energy resources.
- Weather extremes cost money and increase energy usage.

8.2.1.4   Benefits of Technology

On a positive note, the diagram also shows the benefits of *energy technology and policies* and of *environmental technology and policies*. Energy efficiency technology for industrial processes and products (such as appliances or hybrid cars) decreases the *rate of energy consumption*. Advances in extractive and renewable energy technology (such as fracking, efficient solar/wind power, and new renewable energy) increase *energy available*. *Environmental technology and policies* decrease pollution from industry and individuals, and thus reduce *damage to the environment*.

Summary:

- To some extent, technology can reduce environmental damage and increase energy available.

8.2.1.5 Delays

In the diagram, growth of the *global economy* is indeed constrained by *energy available* and by *damage to the environment*. Yet there is a more subtle aspect that corrects our mental model. As ethicists Bazerman and Tenbrunsel point out, we have a "tendency to ignore the future consequences of our actions" (Bazerman and Tenbrunsel, 2011). Thus, our mental model focuses on the present at the expense of the future. Two delays shown in Fig. 8.2 insert a long-term perspective into the picture so that we can grasp the detrimental effects of present actions. These delays mean it takes time to deplete the *energy available* derived from *fossil fuels/uranium* and time to augment renewable *energy available* with *energy technology and policies*. However, the fact that we *are* depleting our limited resources becomes apparent.

Summary:

- System delays hide long-term unintended consequences to environment and economy and must be incorporated in our intervention strategy.

8.2.2 Step 2: Integrating Society with Economy and Environment

The final system diagram in Fig. 8.3 (subsequently referred to as "the system diagram") incorporates population dynamics and other societal interdependencies. This diagram challenges our beliefs about growth. In the previous step, we saw that a rise in energy consumption depletes fossil fuel, heightens environmental damage, and harms the economy. In this step we will see how population growth, increased energy consumption, and economic growth affect the food, water, and economy that keep us alive. Step 2 also integrates *sustainable happiness and well-being* as a complement to personal gratification; this addition will ease our transition to sustainability.

In the diagram, the four elements in large bold print highlight the limiting factors from Chapter 7 – energy supply (primarily *fossil fuels/uranium*), water supply (from underground *aquifers* and *surface water*), food supply (*food-producing land and fisheries*), and *forests* (when converted into *food-producing land*).

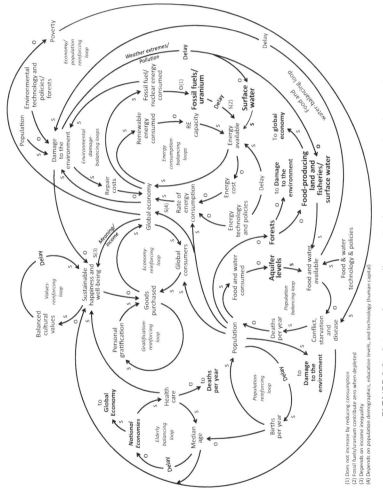

FIGURE 8.3 Integrated system diagram (society, economy, and environment).

[1] Does not increase by reducing consumption
[2] Fossil fuels/uranium contribute zero when depleted
[3] Depends on income inequality
[4] Depends on population demographics, education levels, and technology (human capital)

8.2.2.1 Population Dynamics

To integrate society with economy and environment we first examine the effects of population changes. The *population-reinforcing loop* on the bottom left of Fig. 8.3 depicts population growth or decay. Although global population today is increasing, individual nations may experience either growth or decay. Of course, population cannot increase indefinitely; we will introduce a *population-balancing loop* shortly.

To trace growth in the *population-reinforcing loop*, consider the case in which *births per year* increase *population*. Children mature and have their own children thus raising *births per year* and continuing the growth cycle. Next, consider the case in which *births per year* fall below the replacement rate as in Japan and other nations. Rather than a growth loop, this loop becomes one of decay; *population* can drop when there are fewer people to have children.

Summary:

- More people produce more people. Population decline may produce more decline, depending on birth rate.

8.2.2.2 Population and the Economy

Next, we consider the large *economy/population-reinforcing loop* in the right half of the diagram. Recall from Chapter 6 that increased *population* in part contributes to growth of the *global economy* and that *population* decrease diminishes it. A declining *global economy* can increase *poverty* that then increases *conflict, starvation, and disease*. These links also hold when the *global economy* grows: *poverty* and *conflict, starvation and disease* decrease. Continuing around this long loop, increased *conflict, starvation, and disease* increase *deaths per year*, which is part of the *population-balancing loop*. Finally, more *deaths per year* reduce *population*.

In this loop, economic decline feeds on itself by reducing *population* through *poverty*. The strength of this relationship is dampened since economic growth is not entirely dependent on population. Earlier chapters noted that other factors such as efficiency and innovation brought about by education and technology have contributed to economic growth since World War II. These factors appear as a footnote rather than as a direct link in the diagram.

Population demographics affect the economy in another way. As *births per year* decrease, *median age* rises. Higher *median age* means fewer working age people to care for or support the elderly. In the diagram, after a delay, increased *median age* can burden *national economies* and decrease the *global economy*. There are counterbalancing conditions in this relationship. First, if population drops because parents choose to have fewer children, theoretically these parents can better provide education for their children, who then have more potential to contribute to national economies. Second, in some cultures, elders live with younger generations and affect economies less directly.

Let us digress to make an important point. It is easy to conclude that a nation will experience economic tensions from an aging population, but we must look more deeply. Today's social structures and norms evolved from the past when populations were growing, when there were many more youth than elderly, and when life expectancy was lower. Today, we outlast our retirement savings and strain government programs and young workers. In addition, some cultures discount potential contributions of older workers. These factors do not mean that people must stay employed until they die; rather they say that we must adapt retirement rules and structures to our increased longevity. Thus, the issue is not confined to *median age*, but extends to other social structures that support the nonworking population. Considering both structure *and* aging populations as root causes will help us rethink possible solutions. For simplicity, the system diagram does not address the entire "nonproductivity" issue.

Summary:

- Population growth contributes to economic growth; population decline can burden economies as median age increases.
- Lower birth rates increase median age and create an aging population.
- Economic downturn adds to poverty and can lead to life-threatening situations that ultimately reduce population.

8.2.2.3 Median Age, Health Care, and the Economy

We have already noted that a change in *median age* affects *national economies* by altering the productive workforce. Adding the connection between *national economies* and *health care*, and between *health care* and *median age*, creates the *elderly balancing loop*. Here, healthy *national economies* can improve a nation's *health care*, whether through government subsidy or by giving people the means to afford it; on the other hand, a depressed economy diminishes *health care*. Then, more *health care* increases life expectancy and raises *median age* (depending on *births per year*). Now we are back to the point at which a higher *median age* depresses *national economies*. Note the link between *national economies* and the *global economy*. Because countries of the world are interdependent, economic success or failure in one country affects the economies of all.

Summary:

- National economic growth improves health care but can add to aging population issues.
- Each country's economic success or failure affects the global economy.

8.2.2.4 Health Care, Population, and Happiness

In the diagram, the effects of increased *health care* can oppose one another. By increasing longevity, better *health care raises median age* that can ultimately *diminish sustainable happiness and well-being* as the youthful population takes on more elder care and the older population experiences more health issues.

However, more *health care* also increases *population* by decreasing *deaths per year*. In this case, if infant mortality drops faster than elder mortality, more youth *reduce median age*, which *promotes sustainable happiness and well-being*. Such conflicting relationships emphasize multiple possible outcomes and the need for balanced solutions. Narrow and singular solutions may have unintended outcomes.

Summary:

- Health care is a double-edged sword for sustainable happiness and well-being; solutions must accommodate both short-term and long-term effects.

8.2.2.5   Population and Limits to Food and Water

Next, note the *population-balancing loop* connected to the *population-reinforcing loop* at the bottom of the diagram. As *population* increases, more people raise the amount of *food and water consumed*, which erodes *aquifer levels*. Since aquifers are a limited resource, lower *aquifer levels* reduce *food and water available*, which stimulates competition and theft, and generates *conflict, starvation and disease*. A rise in these conditions reduces *sustainable happiness and well-being*; it also increases *deaths per year* and decreases *population* (*deaths per year* are also influenced by longevity, which is not shown). On the other hand, *food and water technology and policies* can increase *food and water available* and ultimately increase *population*.

Summary:

- More people consume more food and deplete aquifer levels.
- Food and water shortages create life-threatening situations, increase death rates, and reduce population.

8.2.2.6   Population and Damage to the Environment

Next we connect the environment to the social side of the equation at the bottom of the diagram. *Population* growth increases *damage to the environment* in the form of pollution (GHG, air pollution, municipal solid waste, radioactive waste, industrial waste, agricultural waste, and human waste).

In the upper right, increasing *damage to the environment* depresses *sustainable happiness and well-being*. In this case, pollutants weaken our physical and mental health (see Chapter 6). Additionally, environmentally induced weather extremes reduce labor productivity, particularly where searing heat constrains work on outdoor jobs such as construction (Begley, 2014).

Damage to the environment also decreases *food-producing land and fisheries* that, in turn, shrinks *food and water available* – potentially below what *technology and policies* can offset. In this case, *weather extremes* (caused by GHG) distress agriculture and fish habitats and create *surface water* shortages in some regions; other wastes diminish productivity of *food-producing land and fisheries*.

Although *food and water technology and policies* increases *food and water available*, traditional farming releases pollutants from chemical fertilizers, causes erosion, and overuses topsoil; these factors intensify *damage to the environment.* These relationships create the *food and water-balancing loop* in which environmental damage constrains *food and water available* and food production increases *damage to the environment.*

When forests are converted to *food-producing land,* higher *population* means fewer *forests.* Deforestation generates GHGs, thus increasing *damage to the environment.* In addition, when *forests* are removed, so are their abilities to cleanse CO_2 from the atmosphere. Although harvesting forests creates economic activity, for clarity, the diagram omits the link.

Damage to the environment has huge systemic repercussions. The system diagram reflects environmentalists' warnings that damage not only manifests as "electrical blackouts and fuel crises" (Heinberg, 2009) but also as unprecedented levels of "warfare, genocide, starvation, disease epidemics, and collapses of societies" (Diamond, 2005).

Summary:

- More people create more environmental damage and reduce forests.
- Environmental damage reduces food and water supplies.
- Producing more food and water and reducing forests cause environmental damage.

8.2.2.7   *Cultural Values, Sustainable Happiness, and Well-Being*

At the top of the diagram, long-term *sustainable happiness and well-being* replaces *personal security* in the mental model. Now a healthy *global economy* enhances *sustainable happiness and well-being* in the form of *income* and *meaning.* In contrast, economic decline diminishes *sustainable happiness and well-being.* A dark side of economic growth is footnoted in the diagram. Depending on a nation's governance and culture, a flourishing economy may amplify income inequality and thus diminish happiness.

The *values-reinforcing loop* at the top of the diagram counterbalances the short-term *gratification-reinforcing loop.* This new loop incorporates *sustainable happiness and well-being* and *balanced cultural values* which include belonging, relationships, meaningful work, and significance. As these sources of internal well-being increase, reliance on *goods purchased* tends to drop, which weakens the *gratification-reinforcing loop* (and can inhibit the *economy-reinforcing loop*).

Again, we must be clear that not everyone follows the instant gratification path to happiness, but that it is a dominant driver of economic growth. This short-term perspective characterizes today's trend toward Western individualism and consumerism. The values path is more representative of Eastern values that promote collectivism and search for meaning. Together they create a beneficial balance between economy and society.

Summary:

- Economic growth makes us happy (or unhappy when income inequality is high).
- Values that embrace internal satisfaction and de-emphasize instant gratification can smooth the bumps of economic decline that results from fewer *goods purchased.*

8.2.2.8 Delays

Fig. 8.3 adds five additional decades-long delays into the final system. Again, introducing this long-term perspective reveals the delayed effects of present actions. First, the *values-reinforcing loop* has a built-in delay; it takes time to alter cultural beliefs and values. The second and third delays appear between *damage to the environment* and *food and water available* in both directions and once more allow us to ignore damage as it accumulates. Fourth, there is a long delay in the *population-reinforcing loop* between *population* and *births per year.* This delay represents the time required for births per year to decrease. Because it takes a generation for girls to reach childbearing age, the present population will continue to have children even after the *birthrate* has substantially decreased. Finally, there is a delay between a change in *median age* and its effects on *national economies.* Even as people retire, they may not immediately require external support.

Summary:

- Extended system delays add complexity to the system since the effects of present actions may not be seen for decades. We must consider these delays in our intervention strategy and we must have patience.

8.3 MENTAL MODEL AND INTEGRATED SYSTEM COMPARED

What makes the integrated system diagram in Fig. 8.3 different from our mental model? And why are we not experiencing consequences implied by the system diagram? The answers are twofold.

First, the world today is energized by unrestrained growth – growth of economy, personal gratification, energy consumption, and pollution. Like our mental model, the system diagram reflects growth; however, it also adds delays and limits that transform growth into decay. Built-in delays in the balancing loops that constrain *energy consumption, environmental damage, food and water,* and *population* have rendered them feeble in the present; thus we are not yet suffering detrimental effects of continuous growth. Because we have not experienced catastrophic consequences, and because the worst of these consequences may be beyond our lifetimes, our mental model discounts these limits. Second, the *values-reinforcing loop* is relatively weak; today's predominant beliefs emphasize one side of human nature and overpower the benefits of the values loop. Addiction to the instant pleasure expressed in the *gratification-reinforcing loop* has spread like a virus around the world.

8.4 A VIDEO OF THE FUTURE

By incorporating time delays and balancing loops in the system diagram, our perspective is no longer a quick snapshot of current events. We now see a disturbing video of how the future will unfold if current trends persist. The following paragraphs contrast two possible futures using population as an independent variable: One of the two futures is instigated by increased population, and the other by decreased population.

TABLE 8.1 World Population Increases (2015–2030)

Element	Change	Projected 2015 level	Potential 2030 level	Percent change
Population (Billion people)	Up	7.25	8.315	15%
Global economy ($1M GDP, PPP)	Up	96,750	150,740	56%
CO_2 emissions (Million tons carbon)	Up	33,800	41,500	23%
Municipal solid waste (k tons/day)	Up	4,062	7,588	87%
Fossil fuel consumed (MTOE)	Up	11,333	13,904	23%
Forest (Million hectares)	Down	78 m hectares lost		
Aquifer levels (cu km /year)	Down	3583 cu km lost (size of Lake Huron)		

CO_2 emissions and fossil fuel energy consumption from EIA (2013); GDP estimates from PricewaterhouseCoopers (2013); MSW data extrapolated from Hoornweg and Bhada-Tata (2012) at ~5% rate increase per year. Groundwater depletion rate from 2000 to 2008 of 23.9 cu km/year from Konikow (2013) assumed for 15 years; deforestation rate of 5.2 hectares per year from FAO (2012). Corn, rice, grain production ~1.22% increase per year from 2010 to 2012 extrapolated from FAOSTAT (2013). Arable land data from FAOSTAT (2013) extrapolated from ~0.081% decline/year from 2005 to 2010.

8.4.1   Future 1: Population Increases

If nothing changes and we fast-forward several decades, we will see balancing loops gain strength as resources approach their limits, pollution becomes severe, and economies reach their capacity to assimilate nonproductive citizens. A huge and aging population consumes more energy, food, and water than Earth and technology can supply. Weather extremes intensify. Economic growth declines. Depending on technology, innovation and education, the drop could promote depression, unemployment, and shortages. Conflict, disease and death increase. Sustainable happiness and well-being plummet, even as personal gratification weakens from economic decline.

Let us embellish this case with a few numbers. While we want to avoid fortune telling or designing a computer model on the fly, we will use the system diagram and trend data from earlier chapters to see how various elements behave. Suppose that world population grows by one billion (the approximate growth between 2015 and 2030).[1] Table 8.1 shows how other factors could change.

In this future, we have used up substantial amounts of groundwater[2] and fossil fuels. Energy costs rapidly increase. To provide the same food security

---

1. US Census Bureau (2013). Estimated population: 2015 = 7.25 billion; 2029 = 8.25 billion; in 2030 = 8.32 billion.
2. Decreased aquifer levels are about the size of Lake Huron (3540 cu km), the world's sixth largest lake by volume.

with the same amount of land and water, productivity of land and fisheries must accommodate a 15% population growth.

Sustainable development advocates Michael Carley and Ian Christie in the United Kingdom reinforce this bleak view. Our mental model compares to their "technocentric" perspective which risks "unsustainable disruption of ecosystems." They propose an alternative in which stewardship enhances human development and maintains our environment for future generations (Carley and Christie, 2000). Interventions to achieve sustainability proposed in Chapter 9 echo this need for stewardship.

8.4.2   Future 2: Population Decreases

Now let us consider an alternative future some 30–40 years from now. Suppose that the birth rate has dropped well below the replacement rate for at least a generation and global population is decreasing. Although we have neither the data nor the experience to make precise predictions, we will use the system diagram to suggest what might happen.

Decreased population reduces consumption of food, water, and energy, thus stretching the availability of groundwater and fossil fuel and giving more time for renewable energy to mature. Although environmental damage still affects us, fewer people mean that accumulation of damage and life-threatening conditions decrease. Greenhouse gas emissions drop and keep us below the extreme danger threshold for global warming.

On the other hand, there are fewer consumers and workers to contribute to the global economy; median age increases and elder care issues intensify. The economy may decline quickly or gradually, depending on technology, innovation, and education. In the event of extreme economic decline, poverty and life-threatening conditions increase. Happiness and well-being may drop substantially with economic decline, or may stabilize depending on whether cultural shift toward sustainable sources of happiness has occurred.

Both of these cases suggest that we are poised for a rough transition – perhaps over several generations. Both diminish happiness and eventually depress the economy. The first, however, offers no hope for the future; in this case, the system will operate until it exceeds all limits and society crashes. The second puts us on the right path; it improves the environment and reduces the immediate stress of energy depletion and population growth. Here, we have hope of reaching a state of dynamic balance in which population and lifestyles have adapted to what Earth's resources can support. These cases illustrate that depending on what we do in the coming decades, we can either reach *limits-to-growth* thresholds gradually and adapt to change, or hit them abruptly and rock society's very foundations.

We could have used independent variables other than population to see how change affects system behavior. For example, we could try miracles in technology or a drastic reduction of pollution or of energy consumption. In any event, we still must live within Earth's carrying capacity and consider the finite resources that contribute to that capacity. For fun, you may wish to explore other "what if" scenarios.

8.5   IMPERFECTIONS AND LESSONS

Before we proceed, let us first note that the system diagram is not a panacea; its interactions are simplistic. Instead, it is a tool that deepens our insights. Many other influences and relationships could be shown including the effects of national governance on happiness; the influence of culture and economy on health; the consequences of a shift in the balance of power among nations or from nations to transnational corporations; education; the benefits of information technology; and the economic influence of social welfare policies. Furthermore, the diagram does not trace all possible loops; you can discover some of your own.

Yet, in spite of its omissions, the diagram has advantages. First, while it reflects global interdependencies for the most part, it is valid at a lower level of analysis – say, at a national, community, or individual level – if we consider that these smaller units influence overall system behavior. Contributions to population growth, economic growth, or environmental damage can be traced at these levels, although their effects may differ from the global aggregate. Second, the system diagram clarifies our view of the future and allows us to assess what can happen when we apply solutions.

By incorporating current growth trends into the system diagram (see Future 1), we are struck by three current areas of system imbalance: (1) a focus on the present at the expense of the future (short term vs. long term); (2) a preference for self-interest over community interest (individual vs. community); and (3) unequal emphasis among economy, environment, and society. These three areas bring us squarely into the territory of ethical choices and the dilemmas we will face.

8.5.1   Short Term versus Long Term

In the first area, ethicist Rushworth Kidder describes the choice between *now* and *then* as one of the big four "right versus right" paradigms.[3] Our mental model strongly favors the present and dismisses future generations. Still, while sustainability requires that we attend to the future, we cannot disregard the present. Before we evangelize about environmental or population concerns, we must also consider our present quality of life. Kidder laments our less than spectacular involvement with long-term goals of environmentalism, but suggests that "attention to the short term is so important, and so natural a human response, that we sometimes forget to give it the intellectual attention it deserves" (Kidder, 1993). We must therefore strike a balance between present and future.

---

3. Kidder (1993). The second dilemma pits self against community as described in the next section. Kidder's other two dilemmas are justice versus mercy and truth versus loyalty.

8.5.2 Self-Interest versus Community Interest

In the second area, Kidder suggests that the "assertions of individualism and the claims of the community" create the most potent "right versus right" dilemma of all (Kidder, 1993). Over a century ago, biologist Thomas Huxley described this millennia-old debate as "the conflict between man as an individual and man as a social being, the antagonism of selfishness and altruism" (Huxley and Huxley, 1947). We know it well! Today, this tension between self and society is resolving in favor of personal gratification over community welfare. Again, we must not give our own well-being greater weight than the well-being of society – or vice versa. We must balance them rather than choose between them.

8.5.3 Economy, Environment, and Society

In the third area of imbalance, the system diagram demonstrates that disproportionate emphasis on any one facet of sustainability (e.g., the economy) eventually collapses the system. In systems thinking terms, this collapse occurs when the inherent limits from balancing loops halt growth-oriented reinforcing loops, that is, the system applies the brakes. The diagram also illustrates that sustainable happiness and well-being emerge from a synergy among economy, environment, and society – the basic premise of sustainability.

Like the Tree of Souls in *Avatar* and the Gaia hypothesis, sustainability depends upon integration and interdependence – the smooth interaction of its many parts. It is affected by the economic, environmental, and societal decisions of every earthly soul. So, given these fundamental characteristics, what actions can we take to achieve sustainability? The answer has been sprinkled throughout this chapter and can be summed up in a word: "balance."

The need for balance becomes obvious when we trace paths of influence in the system diagram created in this chapter. Using this diagram to compare two alternative futures with opposite population dynamics, we saw how the system reacts. Regardless of whether population increases or decreases, there will be a transition period in which the economy declines and happiness decreases. The difference between the two futures is that in the first case, society eventually deteriorates and in the second case, society ultimately adjusts to a new state of well-being – a state in which the system has found a dynamic balance.

Because everything is connected and because humans are involved, introducing fixes in areas that *seem* most significant in the short term can actually cause severe and unintended outcomes over time. We simply cannot seek personal gratification and individual survival in the present above all else. We cannot grow the economy while ignoring the environment and society and expect the system to achieve sustainability. We cannot increase population and expect Earth's resources to support us. On the other hand, because there are long delays in the system, current future-oriented solutions may *seem* to have no effect. Thus we must be patient and expect gradual progress rather than an immediate transformation.

With eyes fixed on the future, we must consider how to make changes in the context of the whole system and in the presence of other actions. Chapter 9 takes on this very challenge of how and where to intervene. It is, truly, a matter of balance.

REFERENCES

Bazerman, M., Tenbrunsel, A., 2011. Blind Spots: Why We Fail To Do What's Right and What To Do About It. Princeton University Press, Princeton, NJ.

Begley, S., 2014. U.S. to face multibillion-dollar bill from climate change: report. Reuters. Retrieved from <http://news.yahoo.com/u-face-multibillion-dollar-bill-climate-change-report-041353090.html>.

Carley, M., Christie, I., 2000. Managing Sustainable Development, second ed. Earthscan, London.

Diamond, J., 2005. Collapse: How Societies Choose to Fail or Succeed. Penguin, New York.

EIA, 2013. International Energy Outlook 2013. U.S. Energy Information Administration. Retrieved from <http://www.eia.gov/forecasts/ieo/world.cfm>.

FAO, 2012, December 17. Global forest resources assessment 2010. Food and Agriculture Organization of the United Nations. Retrieved from <http://www.fao.org/forestry/fra/fra2010/en/>.

FAOSTAT, 2013. FAO Statistical Yearbook 2013. Food and Agriculture Organization of the United Nations. Retrieved from <http://faostat.fao.org/>.

Heinberg, R., 2009. Searching for a miracle: net energy limits & the fate of industrial society. International Forum on Globalization and the Post Carbon Institute. Retrieved from <http://www.postcarbon.org/new-site-files/Reports/Searching_for_a_Miracle_web10nov09.pdf>.

Hoornweg, D., Bhada-Tata, P., 2012. What a waste: a global review of solid waste management. Urban development series; knowledge papers no. 15. World Bank, Washington. Retrieved from <http://documents.worldbank.org/curated/en/2012/03/16537275/waste-global-review-solid-waste-management>.

Huxley, T.H., Huxley, J., 1947. Evolution and ethics: 1893–1947. The Pilot Press, London.

Kidder, R., 1993. How Good People Make Tough Choices: Resolving the Dilemmas of Ethical Living. William Morrow & Co, New York.

Konikow, L., 2013. Groundwater depletion in the United States (1900–2008). Scientific Investigations Report 2013–5079. US Geological Survey. Retrieved from <http://pubs.usgs.gov/sir/2013/5079/SIR2013-5079.pdf>.

Lovelock, J., 2000. Gaia: The Practical Science of Planetary Medicine. Oxford University Press, Oxford.

Lovelock, J., 2009. The Vanishing Face of Gaia: A Final Warning. Basic Books, New York.

PricewaterhouseCoopers, 2013, January. World in 2050 the BRICs and beyond: prospects, challenges and opportunities. Retrieved from <http://www.pwc.com/en_GX/gx/world-2050/assets/pwc-world-in-2050-report-january-2013.pdf>.

Siegel, D., 2012. Pocket Guide to Interpersonal Neurobiology. W. W. Norton, New York.

US Census Bureau, 2013. World population: total midyear population for the world: 1950–2050. Retrieved from <http://www.census.gov/population/international/data/worldpop/table_population.php>.

Vernadsky, V., 1998/1926. The Biosphere. Complete Annotated ed, translated by D. Langmuir, annotated by M.A.S. McMenamin. Springer Verlag, New York.

Chapter 9

Creating Balance: Effective Interventions

Give me a lever long enough and I will move the world.

– Archimedes[1]

...social systems seem to have a few sensitive influence points through which be-havior can be changed. These high-influence points are not where most people expect. ... the chances are great that a person guided by intuition and judgment will alter the system in the wrong direction.

– Jay W. Forrester (Forrester, 1971)

In June 1859, French acrobat Jean-François Gravelet, familiarly known as the Great Blondin, amazed audiences by walking a tightrope that was suspended 160 ft. above the Niagara River between the United States and Canada (see Fig. 9.1). Spectators paid the incredible sum of 25 cents for the 20 minute show, undoubtedly holding their breaths when Gravelet stopped mid-walk, lay down on the rope, and then continued across. His feat epitomizes exquisite balance; without it, consequences would be deadly.

Just like Gravelet, sustainability requires balance – balance between present and future; balance among economy, environment, and society; balance between national sovereignty and world collaboration; balance between self and community.

This book began with the idea that to achieve sustainability, we must first view the Earth and our role on it as inseparable parts of a whole; we belong to the same system. It described how this system behaves according to internal interdependencies and relationships. In previous chapters, we identified limits to factors that we have long considered irrelevant or infinite. We noted global trends whose delayed effects are ready to materialize in the not-so-distant future and we questioned whether the short-term, self-focused orientation of human nature is compatible with long-term sustainability. Finally, we recognized that this system has slipped out of balance and is headed down a path of decay. If we are to redirect its path, we must adjust the relationships within.

---

1. Geymonat (2011). Archimedes' famous expression is also cited as "give me a place to stand and I will move the world."

K.L. Higgins: Economic Growth and Sustainability. http://dx.doi.org/10.1016/B978-0-12-802204-7.00009-8

FIGURE 9.1 Balance and The Great Blondin. *Photo of Gravelet (The Great Blondin) crossing Niagara Falls with manager Harry Colcord on his back. Retrieved from <http://en.wikipedia.org/wiki/File:Charles.Blondin.jpg>.*

But which relationships? How can we achieve the balance needed to avoid our own fall? Many facts, projections, anxieties, and insights have led us to this chapter; now it is time to gather our optimism and face the challenge of returning our system to a state of balance. We already have an ideal way to address this goal. The systems thinking approach that placed sustainability issues in their larger context can help diagnose dysfunctional behavior and identify critical areas where small changes can make big differences.

9.1 SUSTAINABILITY SOLUTIONS: SYSTEMIC OR SUBOPTIMAL?

It is easy to compartmentalize the elements in our system and to prescribe fixes for specific problems. For example, when national economies sag, policymakers lower interest rates, offer tax cuts, or raise the minimum wage to stimulate economic growth. Such suboptimal and short-lived solutions can have relatively immediate and unintended effects, such as inflation or unemployment. But, as we might deduce from the system diagram in Chapter 8, narrowly directed actions also have long-term repercussions. Actions that encourage economic growth without compensating for environmental damage eventually produce economic downturn, particularly when cleanup costs multiply or when excess greenhouse gases create climate change. Actions that promote population growth accelerate environmental damage that ultimately depletes resources and harms that population.

9.1.1 Leverage

It is indeed demanding to find areas where we can effectively shift a system's behavior without creating adverse effects down the line and where we can make

FIGURE 9.2 Archimedes and the power of leverage. *Engraving from* Mechanic's Magazine, *1824, vol 2. Knight & Lacey, London; reprinted with permission of Kislak Center for Special Collections, Rare Books & Manuscripts, University of Pennsylvania Libraries.*

significant and lasting improvements. Leverage, a powerful tool of systems thinking, is well suited to the task. The idea of leverage has been around at least since the third century BCE when Greek mathematician Archimedes described the principles of a lever. The old engraving in Fig. 9.2 shows Archimedes moving the world and personifies the incredible power of leverage. Just as we rely on leveraging devices such as wheelbarrows or crowbars that amplify the force needed to lift a heavy load, so can we exercise leverage to solve difficult problems.

9.1.2   Generic Types of Intervention

To identify high payoff, high leverage areas for our particular system, we turn to the insights of noted systems thinker Donella Meadows (2008). Table 9.1 summarizes her 12 generic types of intervention (she calls them "leverage points") that can alter a system's behavior. These types of intervention modify underlying assumptions or adjust system components in some way.

From these generic types of intervention, we derived eight high-leverage areas that will move us toward sustainability. Before discussing these areas and their related actions, we first share the approach we developed to identify them so that others can generate their own solutions.

9.2   ANALYTIC APPROACH

It takes several steps to find areas of intervention for a system. Each step considers short-term and long-term implications as well as effects on individuals and on society. Recognize, however, that it is not always possible to have positive

TABLE 9.1 Generic Types of Intervention

Order of effectiveness	Generic types of intervention in a system	Description of intervention
1	Transcending paradigms	Develop the power to change values and adapt to new paradigms at will (e.g., be open to other "truths")
2	Paradigms	Alter the mindset out of which system goals, structure, rules, delays, and parameter arise (e.g., reorient mental models and beliefs about how the world works)
3	Goal of the system	Change the purpose or function of the system (e.g., revise business profit orientation, augment instant gratification)
4	Self-organization	Incorporate the ability to add, change, or evolve system structure (e.g., alter national cultures by encouraging variability and diversity)
5	Rules of the system	Change the rules of the system (e.g., refine incentives, punishments, constraints, constitutions)
6	Information flows	Change the structure of who does and who does not have access to information (e.g., provide information about individual resource usage and pollution)
7	Reinforcing feedback loops	Change the strength of growth and decay loops (e.g., stop epidemics, increase interest rate of savings accounts)
8	Balancing feedback loops	Change the strength of balancing loops relative to the effect they are trying to correct (e.g., increase exercise and good nutrition to lose weight, apply pollution taxes to capture cost of damage)
9	Delays	Change the length of time relative to the rate of system change (e.g., reduce time to build a power plant; increase time to deplete nonrenewable resources; reduce time to implement policies)

TABLE 9.1 Generic Types of Intervention *(cont.)*

Order of effectiveness	Generic types of intervention in a system	Description of intervention
10	The structure of material stocks and flows	Change physical parts of the system and how they interconnect (e.g., alter structures such as health care that determine the median age of population)
11	Size of buffers	Change the size of buffers relative to their flows (e.g., reduce inventory, increase breeding population for endangered species)
12	Numbers	Alter constants and parameters such as subsidies, taxes, standards (e.g., reduce national debt or pollution; alter minimum wage)

This table was derived from discussion of interventions in Meadows (2008); some examples provided by author.

outcomes everywhere. Compromise will be necessary to find balance. Steps to identify these areas are as follows:

- *Find candidate areas*: Search the system diagram for areas that fit any of the 12 generic types of intervention and list them as candidates for further assessment. For example, the second most effective type of intervention recommends altering the mindset or the paradigms from which the system arises, that is, beliefs that have the largest influence on the system. In our case, the most powerful paradigm is the mental model discussed in Chapters 2 and 3.

- *Validate feasibility*: Next, make certain that possible actions in candidate areas can be implemented, make a positive difference, are reasonable in the present, and have minimal drawbacks in the future. For example, although forcing an *immediate* downturn in the *global economy* (e.g., harsh limits on manufacturing) or an *immediate* reduction in *population* (e.g., stiff penalties for having children) would have rapid results, such actions do more short-term harm than long-term good. Eliminate these areas or hold them as last resorts.

- *Prioritize*: Once the candidate list is pared down, prioritize the remaining areas of intervention and select those that are most effective. Take care here. Although we want to leverage our efforts, we should not discourage activities in low-priority areas if they make a positive contribution without long-term damage. For instance, reducing *damage to the environment* by recycling falls at the bottom of the list (it reduces the "number" or amount of accumulated garbage), but we must not dishearten recyclers. Even though independent and suboptimal actions do not reach the heart of the system and have comparatively little systemic effect, they can raise awareness and influence mental models. Instead, we could pair this less effective effort with a more effective one, such as reducing the source of garbage.

- *Create synergy*: Ensure that the areas of intervention on the list are synergistic, do not compete with one another and push system behavior in the desired direction. Note any exceptions and consider how to compensate for opposing forces.
- *Define actions*: List specific actions in each area of intervention. In our case, two to three actions described possibilities. Resolve cases in which specific actions counteract one another. Resolution may require complementary actions in other areas. For example, to achieve sustainability, we should put more emphasis on reducing birthrate than on growing more food, but both are needed to achieve a smooth transition.
- *Assign dates and responsible parties*: Consider *when* the actions should be accomplished and by *whom*. Because system delays can cause unintended consequences, anticipate the effects of delays in your recommendations. In our case, optimum success comes from adopting a long-term philosophy that encourages consistency, stimulates collaboration, and implements actions in *all* areas of intervention as soon as possible.

9.3   AREAS OF INTERVENTION

Many interventions to achieve sustainability are possible. Some are short-lived and may do greater harm than good, thus we focus on areas that have a better chance of accomplishing long-term change.

Using several generic types of intervention from Table 9.1, we have identified eight areas of intervention. Table 9.2 lists them in order of effectiveness and indicates which intervention types apply. Because the most effective areas change the system's fundamental structure – its mental models, goals, and rules – they are also the most difficult to implement. However, even interventions with lower effectiveness, such as increasing energy costs, can alter system behavior.

Fig. 9.3 locates these eight areas (where intervention has the most influence and highest probability of success) on the system diagram from Chapter 8. They are: (1) mental model, (2) balanced cultural values, (3) energy cost, (4) births per year, (5) median age, (6) environmental technology and policies, (7) energy technology and policies, and (8) food and water technology and policies.

By following any arrow from its origin, we can trace its systemic influence. For example, intervention in area 2 (balanced cultural values) directly affects three major reinforcing loops (values, gratification, and economy loops) and rests at a system nexus where sustainability's three components of economy,

TABLE 9.2 Specific Areas of Intervention versus Types

Area of intervention in the system diagram	Generic type of intervention	Effectiveness ranking	Description of area
1	Mindset or paradigm from which the system arises. Information flows	2, 6	Mental model
2	Goal of the system	3	Balanced cultural values (values reinforcing loop)
2, 3	Rules of the system	5	Balanced cultural values (values reinforcing loop) Energy cost (basis of price structure)
2, 4	Strength of reinforcing loops	7	Balanced cultural values (values, economy and gratification reinforcing loops) Births per year (population and economy/population reinforcing loops)
5, 6, 7, 8	Strength of balancing loops relative to the effect they are trying to correct	8	Median age (elderly balancing loop) Environmental technology and policies (environmental damage balancing loops; food and water balancing loop) Energy technology and policies (energy consumption balancing loops) Food and water technology and policies (population balancing loop; food and water balancing loop)
6, 7, 8	Size of buffers	11	Environmental technology and policies (damage to the environment) Energy technology and policies (rate of energy consumption, energy available) Food and water technology and policies (food and water available)
3, 6	Numbers	12	Energy cost Environmental technology and policies (pollution levels)

See Meadows (2008) for effectiveness rankings: 1 is most effective and 12 is least effective.

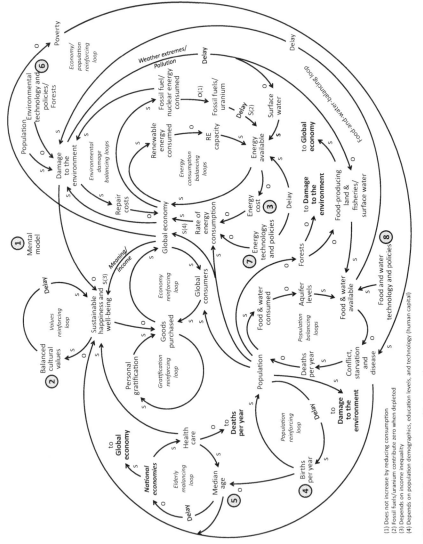

FIGURE 9.3 Eight areas of intervention.

(1) Does not increase by reducing consumption
(2) Fossil fuels/uranium contribute zero when depleted
(3) Depends on income inequality
(4) Depends on population demographics, education levels, and technology (human capital)

society, and environment merge. In addition to counterbalancing today's short-term focus, secondary effects of this intervention quickly spread through the rest of the system.

9.4 PROPOSED ACTIONS

Several actions are possible at each area of intervention. Table 9.3 summarizes a representative set of these actions. Together, they restrain excessive economic growth, reduce environmental damage and reliance on fossil fuels, increase renewable energy sources, develop food and water resources, reduce population, attend to aging, and emphasize sustainable well-being. Note that altering the strength of balancing loops requires changing limiting factors. For example, increased availability of a limited resource weakens the balancing loop; decreased availability strengthens the balancing loop.

9.4.1 Three Levels of Effectiveness

Table 9.3 separates the areas of intervention into three levels depending on their scope and ability to stimulate change: (1) paradigm shifts that form the mental foundation for sustainability; (2) structural changes that alter energy consumption, reduce population, and mitigate the effects of an aging population; and (3) technology-related investments and policies that attend to urgent issues in the present and enable smooth transition to the future. None of these actions explicitly addresses individuals but they all trickle down to individual behaviors. Chapter 13 expands on individual responsibility and self-motivated activities.

9.4.2 Opposing Effects

Outcomes of some actions directly oppose one another or are contrary to the system's goal of long-term sustainability. For example, an obvious way to promote sustainability is to reduce the energy consumption rate and the birth rate. Instead, we have proposed several actions that directly or indirectly encourage births and energy consumption. Why would we do this?

Recall from earlier discussions that we must balance between short-term and long-term effects, particularly when they conflict. Balance requires that we achieve sustainability in ways that consider present society and avoid tragic outcomes.

In the case of birth rate, intervention 8 boosts the availability of food and water to prevent massive starvation (it weakens the population balancing loop and strengthens the population reinforcing loop). These short-term, humane actions accelerate population growth that contradicts sustainability goals. Thus, to balance present with future, this intervention must occur *in concert with* long-term actions that dramatically decrease births per year (intervention 4) to reduce population and actions that deal with the social issues of an aging population (intervention 5). When population growth is at last dampened, the need to increase food and water will diminish.

TABLE 9.3 Intervention Areas and Actions to Achieve Sustainability

Area number	Area of intervention	Actions
Paradigm Shifts		
1	Mental model	Promote system awareness to change the mental model from which the system arises: 1. Arouse moral commitment through fear and inspiration to act as stewards for future generations. 2. Disseminate information in multiple formats and media to broad audiences.
2	Balanced cultural values	Reframe cultural values to strengthen sustainability goals; strengthen the values reinforcing loop; weaken the economy and the gratification reinforcing loops: 1. Champion balanced cultural beliefs about happiness that reduce dependence on buying goods and enhance *sustainable* happiness and well-being. 2. Broaden metrics that define national, business, and individual success beyond GDP, short-term profit/stock prices, and short-term rewards.
Structural Changes		
3	Energy cost	Change the energy cost price structure (rules) to reduce consumption and strengthen the energy consumption balancing loop: 1. Recover the full cost of energy resources (e.g., cost to reduce and remove environmental damage; pay for new infrastructures; compensate for depletion). 2. Prepare citizens and policymakers for increased cost of living.
4	Births per year	Weaken the population reinforcing loop to reduce births per year: 1. Increase education/cultural indoctrination about population overgrowth and birth control. 2. Unite national policies to reduce population. Last resort includes rewards and restrictions on the number of children per family. 3. Increase education levels to counteract effects of population decline on the economy.
5	Median age	Weaken the elderly balancing loop to reduce the burden of the elderly on the economy: 1. Implement social policies that alter retirement age and retirement contributions to ameliorate economic costs of an aging population. 2. Balance the median age among nations through immigration or other policies.

Transition to the Future

6	Environmental technology and policies	Invest in technology and policies that weaken the environmental damage and the food- and water-balancing loops in beneficial ways, and that reduce accumulated damage: 1. Reduce pollution including greenhouse gas, air pollution, municipal solid waste, and industrial and agricultural toxic waste. 2. Repair environment (clean up and reforest).
7	Energy technology and policies	Invest in technology and policies that weaken the renewable energy balancing loop and strengthen the fossil fuel energy consumption balancing loop: 1. Increase renewable primary energy sources that do not damage the environment (increase energy available). Expedite transition from nonrenewable to renewable energy. 2. Reduce the rate of all energy consumption (e.g., efficiency, conservation, recycling). Combine independent and isolated targets for conservation into global goals. The last resort action in this area involves rationing when shortages occur.
8	Food and water technology and policies	Invest in technology and policies to weaken the population and the food and water balancing loops in beneficial ways by increasing the stores of food and water available. Accomplish these actions in concert with reducing population growth. 1. Increase food production in ways that do not damage the environment. 2. Conserve water. 3. Develop new freshwater sources (e.g., desalinization).

In the case of energy consumption, some actions create another apparent contradiction. Intervention 3 increases energy costs and incentivizes conservation, which, together, reduce consumption and environmental damage *but* depress the economy. In systems thinking terms, these actions *strengthen* both the renewable and the nonrenewable energy consumption balancing loops. At the same time, intervention 7 expands renewable energy sources, which encourages consumption and weakens the renewable energy-balancing loop that could limit it. These actions force transition from nonrenewable to renewable energy and enable some level of economic growth.

Do not be disheartened if the system complexities are confusing the first time though – they are neither simple nor evident. However, these examples underscore the criticality of considering both present and future effects of any action in the context of all others. Furthermore, while suggested interventions are by no means comprehensive or perfect, they do represent possibilities. Most importantly, the approach used in this chapter can help evaluate other potential interventions.

The theme of this chapter was balance. It identified eight areas in the system diagram where small efforts generate large results; it then divided these areas into three levels that build the foundation for sustainability, change the structure of current behavior, and plan for transition into the future. Areas of intervention considered both present and future effects, attended to economy, environment and society, and united disparate national policies to achieve coordinated and consistent results. Chapters 10, 11, and 12 provide details for each action in these eight areas.

Because achieving balance in the system is crucial to achieving sustainability, intervention must also be balanced. A cohesive and integrated plan that implements *all* actions in a coordinated way is imperative. Chapter 13 fleshes out this strategy with a discussion of implementation that strikes a balance between self and community.

REFERENCES

Forrester, J., 1971. Counterintuitive behavior of social systems. Technology Review. Alumni Association of the Massachusetts Institute of Technology, Cambridge. Updated March 1995. Retrieved from <http://clexchange.org/ftp/documents/system-dynamics/SD1993-01CounterintuitiveBe.pdf>.

Geymonat, M., 2011. In: Smith, R.A. (Ed.), The Great Archimedes. Baylor University Press, Waco, TX.

Meadows, D.H., 2008. Thinking in Systems: A Primer. Chelsea Green Publishing, White River Junction, VT.

Chapter 10

Pieces of the Puzzle Level I: Paradigm Shifts

Imagination is more important than knowledge. For knowledge is limited, whereas imagination embraces the entire world, stimulating progress, giving birth to evolution.

– Albert Einstein (Einstein, 2009)

Puzzles have entertained us, intrigued us, and soothed us for centuries. In 1760, London engraver John Spilsbury pasted a map on a sheet of wood, cut out individual countries with a fine-toothed saw and sold this first commercial jigsaw puzzle to help children learn geography (McAdam, 2014). He could not have imagined that what began in his workshop as a whim would become a popular pastime so many years later.

Today, jigsaw puzzles come in all sorts of wild shapes and three-dimensional forms – from dolphins and cats to medieval forts and world globes. We could say that sustainability is the largest three-dimensional puzzle of all. To achieve sustainability, one must consider each piece, each action as it relates to all the others; missing just one spoils the big picture.

10.1 SYNERGISTIC PLAN TO ACHIEVE SUSTAINABILITY

In this and the next two chapters, we assemble the puzzle of sustainability in three parts. These parts correspond to the three levels of effective interventions that define our plan (see Chapter 9). Discrete actions proposed in this plan represent individual puzzle pieces. In reviewing these actions, you will note that some are not necessarily new and that their implementation requires more detail and coordination. What is new and unique is that together these actions create a system-wide, integrated strategy that balances present with future, self-interest with community-interest, and economy with environment and society. The advantage of such a strategy is clear, considering that actions are instigated at every corner of the system and that they work synergistically toward a common goal.

K.L. Higgins: Economic Growth and Sustainability. http://dx.doi.org/10.1016/B978-0-12-802204-7.00010-4

FIGURE 10.1 Level I intervention.

10.2 BUILDING THE FOUNDATION

The initial level of intervention lays the foundation of our plan, as symbolized by the partially assembled puzzle of the world in Fig. 10.1. This level incorporates actions for intervention areas 1 and 2 from Chapter 9. These actions change our *mental model* and shape *balanced cultural values* to include meaning, belonging, and other elements of sustainable happiness and well-being.

To aid our discussion, Fig. 10.2 repeats the system diagram from Chapter 9, but highlights these two areas. Effects of actions in area 1 are pervasive. Dashed lines emphasize the direct effects of actions in area 2. This level of intervention is a prerequisite for achieving sustainability and cannot be slighted. It is also the most challenging since it involves a shift of paradigms and beliefs, entails intensive global initiatives, and takes courage to implement.

10.3 AREA 1: MENTAL MODEL

To alter our mental model, we must first understand the dismal consequences of current lifestyles. Economic growth that depends on carbon-based energy, thoughtless consumption, generation of excessive waste, and blind faith in technology cannot continue without destroying our chance for sustainability. Andrew Schmookler underscores the importance of changing the tenets of our mental model:

> *"...we are unlikely to make real progress toward a more balanced social policy*
> *and way of life until we take the first step of recognizing that, however difficult it*
> *may be for us to do so, we need to turn our chariots around" (Schmookler, 1993).*

Bringing this awareness to the large and diverse world audience is intimidating at best. A new mental model will recalibrate power and status for those who deem themselves at the top rung of the economic or social ladder. For some, taking ownership of the damage we are causing will destroy their complaisance and disturb their sense of security. Others will squawk when they realize that proposed actions will impede their lives. And, it will be most difficult to convince skeptics who believe that population growth is not an issue; that technology will solve all problems; that economies *must* grow; and that global warming is either a myth or a conspiracy.

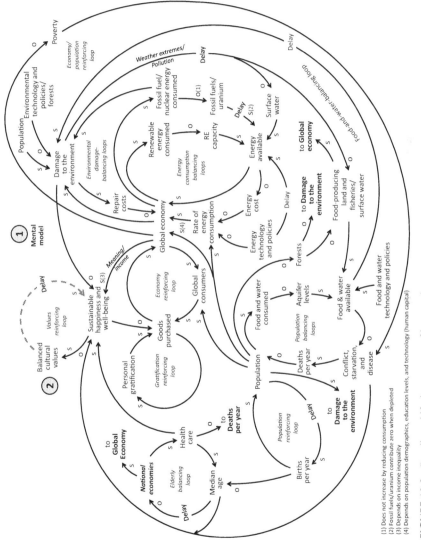

FIGURE 10.2 Paradigm shifts in two areas of intervention.

(1) Does not increase by reducing consumption
(2) Fossil fuels/uranium contribute zero when depleted
(3) Depends on income inequality
(4) Depends on population demographics, education levels, and technology (human capital)

Even when enough of us appreciate the implications of our actions, it will take decades, perhaps several generations, to achieve a sustainable future. As with any force that shakes foundations and alters lives, actions proposed here will not be embraced – at least not initially. We must both push and pull to open minds. We must push with fear of consequences and pull with the inspiration of a better future for our descendants. These actions will: (1) arouse moral commitment; and (2) disseminate information in multiple formats and media. They broaden our perspectives to include long-term effects.

10.3.1 Arouse Moral Commitment Through Fear and Inspiration

To achieve sustainability, we must all share a "moral purpose," says Advisor to Canada's Premier, Michael Fullan (Fullan, 2005). Thus, the initial action in this area of intervention intends to stimulate stewardship and protect the Earth. First, we quantify the damaging consequences of current lifestyles to kindle equal measures of fear and inspiration in individuals, organizations, and nations. This list of consequences should provide compelling examples of environmental and social costs of consumerism, and of economic and population growth.

The system diagram in Fig. 10.2 can help us visualize consequences and discover hope for the future. It also illustrates how current behaviors oppose sustainability and what we can do to promote it. As a start, we know we must conserve resources, reduce material consumption, decrease population, and embrace a different quality of life in the present while we invest in the future.

10.3.2 Disseminate Information in Multiple Formats and Media

Even when we understand sustainability and how to inspire commitment, we still must determine who will coordinate and who will disseminate the information. This task cannot be done haphazardly or independently. It cannot be directed to small niches of society. International groups should coordinate how the message is spread. Although such groups have their own parochial interests and may not initially agree on specific actions, they can appreciate the problem and collaborate on major points.

In addition to coordinating what our message will be and who will communicate it, we must strategize how to reach the general population. Existing information on sustainability is comprehensive and readily available, but much of it is fairly technical and not broadly circulated (e.g., the IPCC report *Climate Change 2013*)(IPCC, 2013). Furthermore, those who believe that humans are not affecting the climate tend to overshadow reported findings with their vigorous objections.

Urgency for change must reach the world's population in various formats and levels of sophistication. We must circulate information in simple but persuasive words and pictures via public media. Art forms such as movies and graphic publications can punctuate rational awareness with emotion.

The messages buried in this information must target both adults whose mental models are well formed and youth whose mental models are still immature.

Through public and private education, we can amplify the sustainability message to the millennial generation (those born between 1982 and 2003), since they will have great influence during the next 25 years. Hope rests with this audience, particularly because they are proficient with electronic and social media, and are acquainted with global issues, different cultures, and environmentally sound practices. Their broad exposure encourages the sort of open-minded decision-making that will be essential.

10.3.3   Ongoing Efforts: Research and Information Dissemination

Discrete efforts to understand, inspire action, and disseminate information are ongoing. On the push side of awareness, many with strong commitments to sustainability are speaking out. Environmentalist Richard Heinberg, for example, urges us "to appreciate that the twentieth century's highly indulgent over-consumptive economic patterns were a one-time only proposition, and cannot be maintained" (Heinberg, 2009).

Many other voices have joined the chorus of those who resonate with this imperative. For example, in 2013 the Intergovernmental Panel on Climate Change (IPCC) made progress in quantifying the effects of GHG emissions (IPCC, 2013). This research, conducted by hundreds of experts from more than 120 countries, is a reasonable starting point for investigation.

Other international groups are examining global warming. For instance, the Global Conference on Global Warming 2014 in Bejing, China provides a forum for information exchange. On a domestic note, the United States Senate recently convened a "newly-formed Climate Action Task Force" to build advocacy for study and action on climate change (York, 2014). Awareness is growing in the general population as well. A recent survey in over 40 US states found that at least 75% of US adults believe global warming is real (Koch, 2014).

Yet, these few cases are not enough. Sustainability requires more proactive, systematic, and far-reaching measures to spread awareness and gain moral commitment. Government, business leaders, and individuals alike must step up to the plate and be heard – soon.

10.4   AREA 2: BALANCED CULTURAL VALUES

Happiness and well-being have long been prime objectives of humankind. Most of our efforts – our scientific advances, economic endeavors, and social policies – intend to add quality to our lives. Intervention here must somehow preserve this goal without compromising future generations. Success depends on our ability to encourage balanced cultural values that cherish the future and alter our perception of well-being, even at some cost to the present. In essence, these values blend Eastern philosophies of collectivism, holism, and discovering meaning in life within a bigger context and Western philosophies that promote individualism and the self-interest implied by consumerism. This value shift is the fulcrum on which the transition to sustainability rests. Rather than reinforcing values that lead to decline, actions here promote environmental and societal health while *stabilizing* or *shrinking* the economy.

Earlier we asked whether we humans can burst our myopic bubbles and regard the future with fresh and caring eyes. Only by viewing sustainability as an integrated system can we appreciate the significance of this question and the magnitude of changes we must make in our beliefs. Diamond affirms that the "crux of success or failure as a society is to know which core values to hold on to, and which ones to discard and replace with new values, when times change" (Diamond, 2005). Others acknowledge that solving economic and environmental issues requires a new value paradigm – one which no longer views economic growth as a measure of happiness and source of international power.[1] It is time to adjust our values.

But recognizing the need to change and accomplishing it are two different animals. Because values are the touchstone for preferences and behaviors, they are difficult to abandon, particularly if they allow us to gratify our desires. We may know that we need to diet, but are hard-pressed to reduce our food intake day in and day out. We may know that exercise improves our health, but it is hard to get out of bed in the morning. So, what will it take to push us over the edge into the fog of change? The impetus, I believe, will come from one of two directions.

One possibility is that given current trends, conditions on Earth will deteriorate as energy, food, and water resources dwindle and the environment becomes more toxic. When these conditions affect survival on a more massive scale than today, we will be forced to reassess priorities and alter behaviors. This path, which Chapter 1 called "overshoot-and-collapse," is a demanding one, peppered with danger and despair. Conflict and competition for everything – from food and water to energy and jobs – will be stumbling blocks. It is the path of imbalance and decay that our grandchildren's children will follow unless we clear the way.

The second possibility suggests that by developing strong alternatives for how we experience happiness and well-being, the system will respond more gently. If we decrease our appetite for personal gratification and enhance the meaning we get from our jobs, relationships, and learning activities, we can reduce the excessive consumption and economic growth that harm the environment and deplete resources. If we have fewer children, we will eventually reduce the pressure on environment and resources. In this option, we gradually adjust to a new norm in which excess population, over-abundance, and economic growth dissipate and where sufficient is enough.

This area of intervention offers two of the many actions that could apply to this second possibility. To intervene successfully, leaders everywhere must:

---

1. See, for example, discussion by sustainability expert AtKisson (2012).

(1) champion cultural beliefs that downplay consumption and encourage sustainable happiness and well-being; and (2) broaden the metrics our society uses to define success. Dashed lines in Fig. 10.2 show that changes to cultural values simultaneously increase happiness and well-being in the long term, and reduce the frenetic drive to purchase goods in the short term.

Engagement in this area is not easy. It requires that leaders first become uncomfortable with our likely future and then respond with gut-level resolve to change. Again, this is an area in which the millennial generation can excel.

10.4.1   Champion Balanced Cultural Beliefs for Sustainable Happiness and Well-Being

Initial actions here promote happiness in ways unrelated to materialism and consumption. They imply deep cultural shifts, particularly in more advanced economies that depend upon abundance. Yet, as huge and difficult and unpleasant as it might be, a value shift is feasible – though it will take decades and constant attention. We *do* have the ability to counterbalance our self-focus; we *can* challenge ourselves by asking: "How much of our traditional consumer values and First World living standard can we afford to retain?" (Diamond, 2005).

This optimism originates from the fact that we have already witnessed massive cultural shifts motivated by circumstances. In the wake of the Industrial Age, for example, productivity rose, economies flourished, and well-being improved. These advances stimulated a cultural transformation that congealed into our present mental model. Schmookler notes two particular values that have surfaced since the nineteenth century: "First, there arose a worship of *success*. This success meant the fulfillment of ambition in the arena of economic competition. And second, there was a tendency to enshrine *wealth* as the essence of value" (Schmookler, 1993). Both values have colored the predominant world view.

Of course the transition from having less to having more is easier than moving in the other direction. Therefore, we must factor in another human characteristic that supports the feasibility of culture shift: adaptability. In the face of dramatic change, we adapt to survive. In a cultural sense, values help us adjust to new situations and promote our fitness for survival (Michod, 1993). Thus, the good news is that because culture is built from social practices in a given milieu, it is, as Cushman says "somewhat moveable" (Cushman, 1995).

One way to change our cultures is to approach *balanced cultural values* directly. In Fig. 10.2 system diagram, *sustainable happiness & well-being* is attached to the *values reinforcing loop*. This loop offsets short-term philosophies and nurtures future-oriented viewpoints that rely on a sense of belonging and meaningful moral purpose. Individuals are capable of committing to something greater than themselves. As their reliance on sustainable happiness increases, they will not be as tempted to fill their internal voids with material goods. The disadvantage of this values loop – but a fact of life nevertheless – is that less consumption weakens economies.

10.4.2 Broaden Society's Metrics for Success

One caveat to achieving such a wholesale value shift is that the social and economic structures supporting our culture must also change. Schmookler's insight is deep in this area. "The transformation of values," he says, "must go hand in hand with a transformation in the economic system" (Schmookler, 1993). In other words, the global economic system – fed by consumers who seek personal gratification, by organizations with short-term profit motives and instant reward policies, and by financial markets that incentivize quarterly returns-on-investment for investors – must change its rules and its metrics for success.

10.4.3 Ongoing Efforts: New Economics and New Metrics

Because it involves cultural values and ingrained structures, success in this area will necessarily occur bit by bit. If only a few well-connected individuals, nations, and companies value long-term results and are successful, others will follow. Examples of trends in this direction offer hope. A recent movement, called "new economics" recognizes the potential to shift our values and suggests that sources of happiness other than economic growth will encourage long-term thinking and conservation. British ecological economist Tim Jackson suggests that beyond a certain point, economic growth hinders rather than improves human well-being (Jackson, 2011). Groups such as the New Economics Foundation and political economists such as David Boyle and Andrew Simms promote a new paradigm for economics that incorporates both societal and environmental costs (Boyle and Simms, 2009).

While these new perspectives make us reconsider our beliefs, other initiatives exemplify how the engines of an economy – the business organizations – are nudging our values away from immediate gratification and toward more lasting success. For example, a recent study on pay practices in the United States found that the percentage of private companies offering long-term incentives in addition to short-term incentives has increased from 35% in 2007 to 61% in 2012 (Sharp, 2012). The "new economics" movement echoes this transition and advocates substituting well-being for GDP as a measure of national success. These efforts show us where to begin; now we must expand their scope.

The two areas of intervention at this first level are demanding and entail collaboration of nations, organizations, and individuals. Associated actions, although highly effective, are not for sissies. They require constancy of purpose and commitment to a long-term outcome; they also necessitate short-term sacrifices and dogged focus to weather social upheaval and criticism. In implementing these actions, it will be important to COMMUNICATE, COMMUNICATE, COMMUNICATE: intentions, plans, possible side-effects, and progress.

The next two chapters tackle the last six areas of intervention. Although they do not share the potency of the intervention described in this chapter, subsequent actions depend upon the new sustainability paradigms and values described here.

REFERENCES

AtKisson, A., 2012. Life beyond growth: alternatives and complements to GDP-measured growth as a framing concept for social progress. Tokyo, Japan: 2012 Annual Survey Report of the Institute for Studies in Happiness, Economy, and Society - ISHES. Retrieved from <isisacademy.com/wp-content/uploads/LifeBeyondGrowth.pdf>.

Boyle, D., Simms, A., 2009. The New Economics: A Bigger Picture. Earthscan, London.

Cushman, P., 1995. Constructing the Self, Constructing America. Addison-Wesley Publishing Co, Reading, MA.

Diamond, J., 2005. Collapse: How Societies Choose to Fail or Succeed. Penguin, New York.

Einstein, A., 2009. Einstein on Cosmic Religion and Other Opinions & Aphorisms. Dover Publications, Mineola, New York.

Fullan, M., 2005. Leadership at the System Level. Corwin, Thousand Oaks, CA.

Heinberg, R., 2009. Searching for a miracle: net energy limits & the fate of industrial society. International Forum on Globalization and the Post Carbon Institute. Retrieved from <http://www.postcarbon.org/new-site-files/Reports/Searching_for_a_Miracle_web10nov09.pdf>.

IPCC, October, 2013. Climate change 2013: The physical science basis summary for policymakers. Intergovernmental Panel on Climate Change. Retrieved from <http://www.climatechange2013.org/images/uploads/WGI_AR5_SPM_brochure.pdf>.

Jackson, T., 2011. Prosperity without Growth: Economics for a Finite Planet. Earthscan, London.

Koch, W., February 9, 2014. Most Americans agree about global warming, but not Congress. Retrieved from <http://peaceworker.org/2014/02/most-americans-agree-about-global-warming-but-not-congress/>.

McAdam, D., 2014. History of jigsaw puzzles. American Jigsaw Puzzle Society. Retrieved from <http://www.jigsaw-puzzle.org/jigsaw-puzzle-history.html>.

Michod, R., 1993. Biology and the origin of values. In: Hechter, M., Nadel, L., Michod, R.E. (Eds.), The Origin of Values. Aldine de Gruyter, Hawthorne, New York, pp. 261–271.

Schmookler, A., 1993. The Illusion of Choice: How the Market Economy Shapes Our Destiny. State University of New York Press, New York.

Sharp, M., February 14, 2012. More private companies offer both short- and long-term incentives to employees. WorldatWork, Scottsdale, AZ. Retrieved from <http://www.worldatwork.org/waw/adimLink?id=58598&from=press4>.

York, B., March 10, 2014. Senate democrats' donor-friendly global warming show. Washington Examiner. Retrieved from <http://washingtonexaminer.com/senate-democrats-donor-friendly-global-warming-show/article/2545354>.

Chapter 11

Pieces of the Puzzle Level II: Structural Changes

> *There are no extra pieces in the universe. Everyone is here because he or she has a place to fill, and every piece must fit itself into the big jigsaw puzzle.*
>
> – Deepak Chopra

In the previous chapter, the first level of intervention proposed to alter mental models and values. Actions at this level are highly effective yet difficult to implement because behaviors are so deeply embedded in our unconscious thought. Although these changes are essential and transform our world view, they do not require us to do anything with this new perspective. This chapter continues with the second level of effective intervention and adds the *doing* part.

11.1 ALTERING FEEDBACK LOOPS

Intervention at the second level builds on the paradigm shifts. Just as the puzzle of the world in Fig. 11.1 is taking shape, so this level adds substance to our sustainability plan. It proposes sweeping policies and revamps economic and social structures to halt unconstrained energy consumption, population growth, and dependence of the elderly on young workers. Because actions here alter system structure (by changing two balancing loops and a reinforcing loop), they are both effective and challenging.

Focal points for change in the system diagram are: *energy cost, births per year,* and *median age*. Fig. 11.2 highlights intervention areas 3, 4, and 5 and uses

FIGURE 11.1 Level II intervention.

K.L. Higgins: Economic Growth and Sustainability. http://dx.doi.org/10.1016/B978-0-12-802204-7.00011-6

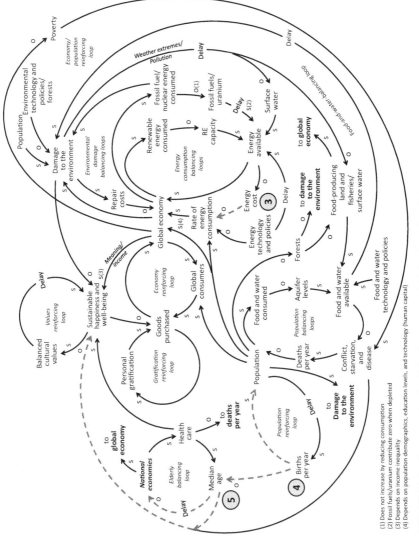

FIGURE 11.2 Structure-related policies for energy cost, birth rate, and median age.

(1) Does not increase by reducing consumption
(2) Fossil fuels/uranium contribute zero when depleted
(3) Depends on income inequality
(4) Depends on population demographics, education levels, and technology (human capital)

dashed lines to emphasize their direct effects on *population, national economies, rate of energy consumption,* and *sustainable happiness and well-being.* Actions in these areas indirectly affect multiple elements including the *global economy, damage to the environment,* and *energy available.*

11.2   AREA 3: ENERGY COST

Although implementing actions in this third area of intervention is more clearcut than in the first two, repercussions are more immediate and the scope is instantaneously broad. Here, we increase the cost of energy to reduce the rate of consumption, fund alternative energy sources, and clean up the environment. Two actions in this area require us to: (1) recover the full cost of energy products and services; and (2) at the same time prepare ourselves for an increased cost of living.

11.2.1   Recover Full Cost of Energy Products and Services

This ambitious action recovers the full costs of energy products and services. It requires a world-united front. For a nation to be provincially proud of its own energy independence does not fly well here. Favorable geography and inherited natural resources for a few do not count toward achieving sustainability for all.

Gasoline, natural gas, and electricity prices have increased over the years, but they do not incorporate hidden costs. To account for full costs, we must isolate the origin, extent, and concentration of harmful byproducts so that energy costs for organizations or individuals can be adjusted according to their "carbon footprints," that is, the greenhouse gas (GHG) and waste they produce. Costs must include long-term expenses associated with environmental damage and with finding alternatives for nonsustainable resources. Parts of this increase should be invested in environmental cleanup and renewable energy technologies. International collaboration ensures that costs are equitable, applied uniformly, and not pumped back into the coffers of nations, organizations, or individuals.

11.2.2   Prepare for Cost of Living Increases

To reverse existing damage and compensate for carbon-based fuel consumption, prices must rise over the next decades to a level that most nations and most people are unwilling to accept without protest. Although energy price hikes are only the tip of the iceberg, they cascade through economies. In the United States for instance, for every 10% increase in energy prices, producer prices increase 1.5%, which reduces their profit and their ability to hire; consumer prices rise 0.7% making *everything* we buy more expensive without a commensurate increase in wages. In other words, individuals and businesses will be squeezed.[1]

---

1. EIA (1998). Had the projected price hikes occurred, US GDP in 2010 would have dropped from 1% to 4.2%.

Policymakers as well as national and organizational leaders must weather criticism for higher costs of living. Nations must keep inflation down. As a result, higher costs of living mean that individuals must cut other expenses rather than expect incomes to increase. Because this situation also depresses world economies, we must all tighten our belts.

Unfortunately, this outcome is nonnegotiable with respect to sustainability; the only variable is *when* price hikes should occur. I submit that it is better to begin immediately and suffer a gradual economic downturn or stagnation that allows us to adapt than to delay action and have the bottom fall out of the global economy. As Benjamin Franklin once said "an ounce of prevention is worth a pound of cure."

11.2.3   Ongoing Efforts: Attempts to Raise Energy Costs

Recovering the full cost of energy products and services is not new. Some companies already advocate higher prices and propose that economic markets undergo "a paradigm shift to Sustainable Capitalism" (Generation Investment Management, 2012). International attempts, such as those promoted by the Kyoto Protocol to reduce GHG emissions, have considered raising energy prices.[2] However, the fact that countries such as the United States and China neither ratified the Kyoto Protocol nor implemented an energy cost recovery policy speaks to the dilemma.

In considering whether to meet Kyoto Protocol targets, the United States projected that coal prices would nearly quadruple and electricity prices would increase by 50% within a decade (EIA, 1998). Besides higher energy costs, other undesirable consequences include substantial economic decline, loss of jobs, and backlash for the politicians who supported cost increases. The United States did not sign up to the Kyoto Protocol for these reasons – another instance of choosing economy and political popularity over environment.

On a more optimistic note, as environmental damage becomes more visible, some countries have acted. Countries in the European Union impose heavy taxes on energy and transportation to reduce air pollutants; they also use a cap-and-trade system to limit CO_2 emissions.[3] Revenue from these taxes accounts for 2.4% of their combined GDP. For individual countries such as Denmark and the Netherlands that number can reach 4% of their GDPs (Eurostat, 2013). Net effects of higher energy costs on these economies is unknown.

In another part of the world, China, with its notorious and noxious air pollution, has begun to address environmental damage. It recently announced new

---

2. The Kyoto Protocol was an international treaty signed in 1997 that intended to obligate industrialized nations to reduce greenhouse gases.

3. Levinson (2007). Cap and trade is a policy that puts a maximum on emissions, but allows flexibility on how companies and governments meet these targets – they can buy, sell, or bank emission credits to use in the future.

emissions standards for heavy-polluting industries and proposes to reduce air pollution at least 30% by the end of 2017 (Associated Press, 2013).

In spite of these measures, harmful emissions from consuming carbon-based energy are still increasing with economic growth. Energy price hikes through taxes or other mechanisms create at least two positive effects. First, some revenue can be dedicated to cleanup, innovation, and efficiency. Second, those who consume energy will work a little harder to conserve it. As personally and politically unpopular as a cost increase is, its benefits will, in the long run, compensate for its negative pressure on the economy. If we take no action we are merely indulging today's desires at the expense of future generations; we are not paying our own way.

11.3   AREA 4: BIRTHS PER YEAR

Managing population growth is a controversial topic and one of the greatest challenges posed by sustainability. All actions in this area must apply *consistently* around the world to achieve a single goal: reduce births per year to decrease world population. Recall from Chapter 7 that Earth's carrying capacity has been estimated at two billion people. Even if this number is pessimistic, we already know that in the long term Earth cannot support our current population of over seven billion.

Actions in this area include: (1) educate people about family planning and culturally indoctrinate them to have fewer children; and (2) unite world policy on population. Changes are not trivial; population control is a value-laden issue that touches religion, ethics, and politics. It boils down to a "conflict between the reproductive rights of the current generation and the survival rights of the next generation" (Brown et al., 2000).

Yet, regardless of the values it touches, the case to halt unbounded population growth is undeniable. By tracing the effects of population growth or decay in the system diagram, as we did for the alternate futures in Chapter 8, we immediately understand its pervasive influence – its effects pop up everywhere like pesky prairie dogs in the Western United States.

11.3.1   Educate and Indoctrinate

Some international agencies vehemently oppose population control as a matter of personal rights; they also suggest there is plenty of land to support more people and that overpopulation is only a problem in urban areas.[4] Others just as fervently promote the need to curb population growth before it reaches the projected nine billion mark in 2042. Education and indoctrination must include a clear and compelling case for reducing the number of births per year.

---

4. Population Research Institute is one particularly vocal agency.

Arguments to promote birth control must transcend current values about having more children. Global values that promote quality of life for all could be helpful; other discussions must be tailored to country, religion, or organization. In addition, people must know the dire and unintended consequences that unrestrained population growth has on environment, economy, and society. None of us should consider ourselves exceptions to the population reduction goal.

11.3.2   Unite National Policies

Although we are one world whose population is increasing, we operate as independent entities. National policies on population control are as diverse as the many charitable causes in the world. Some countries use policies, rewards, and punishments to discourage population growth while other countries are either doing nothing or are encouraging their populations to grow.

To unite our views and our goals, solutions must address both sides of population dynamics: growth *and* decline. Nations whose growing populations threaten their stability must take immediate action to reduce birth rates significantly. They may need to restrict births or reward having fewer children. Then, because of the delay between reducing the birth rate and the leveling off of population, the rest of the world could support population excess.

Thus, nations whose declining populations cannot support their economies could encourage immigration or adoption rather than sending food or money since doing so merely encourages population growth. They could hire unemployed workers from overpopulated areas, improve farming for those who do not have enough to eat, alter emigration and adoption policies, and educate children from nations that cannot support their growing numbers. *Concurrently*, these same nations should *no longer* incentivize having children even though their median age is increasing. Instead, they should emphasize alternative ways to reduce the dependency of older generations. Whatever actions we take, we must take care not to encourage population growth by alleviating consequences at local levels. We must also ensure that displaced populations are treated fairly and humanely.

11.3.3   Increase Education Levels

Chapters 5 and 8 noted that a declining population can depress the economy, although at a slower rate since other factors such as increased median age also contribute to economic decline. Action here increases education levels and allows more people to increase their contributions to national economies, and ultimately to the global economy, thus counteracting some negative economic effects of population decline.

11.3.4   Ongoing Efforts: Policies for Population Control and Population Growth

In some cases, ongoing efforts in this area contradict one another: they support population growth as well as population control. Many nonprofit organizations such as Population Connection,[5] massive charities such as the World Population Foundation,[6] the Gates Foundation,[7] and the Packard Foundation[8] support population control. Together their efforts have had some success. For example, "active use of family-planning techniques in developing countries" rose from 12% in the 1960s to over 60% in 2008 (The Economist, 2008). These successes have already played out as a lower rate of population growth.

Initiatives by these influential groups and individuals can be expanded and supplemented by local endeavors that are tailored to specific circumstances. New actions should be mindful of objections raised by organizations that disapprove of population control, such as the Population Research Institute.[9] They should avoid policies that enforce population stabilization in an unethical manner, for example, forced sterilization or, in some cases, infanticide.

Individual governments have also instituted population control policies. Since the 1980s, Iran's Islamic government has made birth control widely available and has required training in contraception and family planning before couples can get a marriage license. The policy gave more freedom to Iranian women and reinforced a trend to have one or two children at most. The drop in Iran's fertility rate – from about seven children per woman in the 1980s to fewer than two today – was the largest and fastest ever recorded (Weiss and Mostaghim, 2012). This policy points to the link between education and reduced fertility rates (not shown in the system diagram).

China's one-child policy from the 1970s may have "prevented some 400 million births," (Park, 2013); its fertility rate plunged to about 1.7 in 2011 – below the replacement rate (The World Bank, 2014). India has a different story. Rather than mandates, it offers rewards. With a population that will soon exceed China's 1.36 billion and a population density over two and a half times that of China (Mongabay, 2008; International Monetary Fund, 2013), India gives prizes such as televisions, cars, and food processors for those who agree to be sterilized. It also advocates bringing electricity to every village so that people will watch TV rather than make babies (Global Post, 2011).

On the growth side of population policy, low-fertility rate countries such as Australia, Estonia, France, and Germany offer financial and other incentives,

---

5. Population Connection advocates reducing fertility rates and population stabilization. Retrieved from <http://www.populationconnection.org>.

6. The World Population Foundation draws attention to the effects of high birth rates and rapid population growth, with the intent to reduce worldwide poverty. Retrieved from <http://www.factualworld.com/article/World_Population_Foundation>.

7. The Bill & Melinda Gates Foundation brings high-quality contraceptive information, services, and supplies to women in the poorest countries with the intent of expanding voluntary family planning. Retrieved from <http://www.gatesfoundation.org>.

8. The David and Lucile Packard Foundation expands access to sexuality education, voluntary contraception, and other efforts to promote women's reproductive health and to stabilize population growth. Retrieved from <http://www.packard.org/>.

9. The mission of the Population Research Institute is "to end coercive population control and fight the myth of overpopulation which fuels it." Retrieved from <http://pop.org>.

including patriotic duty, for families to have more children.[10] Unfortunately, both Iranian and Chinese governments are rethinking their population control policies; they believe they have reduced their populations too much. Iran's former President Ahmadinejad believes that more Iranians mean a better defense posture and pose a greater threat to the West (Weiss and Mostaghim, 2012). China is trying to remedy the unintended effects of population control. Because it has encouraged having male children, China's gender ratio is lopsided; rather than the norm of 103–107 boys per 100 girls, current statistics are 119 boys per 100 girls. The country is simultaneously facing an aging population that requires care, males who cannot find wives, and fewer young people to work in factories (Lim, 2008). India's situation is again different; at the same time that government policies promote population control, small tribes pay women to bear children – sometimes 15 or more – to defend their land from outsiders (Bhaumik, 2007).

These cases clearly demonstrate the challenges ahead and the need not only to unite population policies, but also to push nations in the direction of decreasing their populations.

11.4   AREA 5: MEDIAN AGE

The fifth area of intervention involves median age and the world trend of an aging population. Chapter 6 described some of the social and economic implications of a society that is growing older. Economically speaking, because elderly people have more health issues and may have left the workforce, younger workers must help with their support either through taxes or out-of-pocket expenses. An aging population also means that with fewer young workers, organizations may experience a shortage of employees. In addition, pressure on the economy combined with the stress on younger generations reduces overall well-being.

As we discussed earlier, some countries constrain population growth while others encourage it to avoid labor shortages, deal with rising health care costs, and add to the dwindling number of taxpayers. This area offers a way to compensate for the economic and social well-being effects caused by the population decline proposed in area 4. Rather than encouraging births in those nations with the most severe issues, it seeks to ameliorate and equalize the global effects of rising median age so that countries are not tempted to ignore or negate actions that reduce population.

Because we share the same world, global remedies must accommodate nations that are experiencing both growth and decline. This intervention has two parts. All countries, regardless of their specific population dynamics, should: (1) revise retirement policies and programs; and (2) adjust their median age to balance with those of other nations. The next section details these actions.

---

10. See various sources listed in <http://www.vhemt.org/bbbounty.htm>.

To gauge the success of policies in this intervention area, we can monitor the elderly dependency ratio (EDR) described in Chapter 6. As median age rises, as it has in the past 15 years, the EDR will increase in most countries whose populations are dropping. Recall that a high EDR means that more elderly must be supported by fewer working-age people, thus the goal of intervention in this area is to decrease EDR.

11.4.1   Implement Retirement-Related Changes

Actions here are controversial since they affect one's subsistence. They support programs that extend retirement age, increase retirement contributions, and/or reduce benefits. These programs must be supplemented by education and health care that enable elderly populations to be a productive part of society. By increasing the retirement age, elderly dependency ratios drop and the number of working age individuals who support the elder population increases.[11]

For populations that are growing, some actions are counterintuitive. Why would countries want to add part of their elder population back into the workforce when they may have a glut of young people who also need jobs? Furthermore, in any country, regardless of whether its population is rising or dropping, why would those about to retire want to delay their pensions? The answer is twofold: (1) to live within the means of one's retirement income and not encumber national economies; and (2) as responsible citizens, to prepare the world for future generations. Both actions require sacrifice in the present.

Over the long term, retirement benefits that exceed what was put aside end up costing nations, states, and communities big bucks and put a greater tax burden on those who are still working. Although there will be a painful adjustment period until their working populations decline, nations will gain a more stable economic foundation by reducing these costs.

11.4.2   Balance Median Age Among Nations

If implemented in all nations, retirement-related policies will decrease the EDR and take stress off economies whose median age is increasing. But what can we do to alleviate the pain for both growing AND declining populations? It is impossible to *eliminate* the pain, but it can be spread around. For example, to achieve comparable median age among nations, relaxing immigration policies allows individuals from countries whose populations are growing to fill the void in countries whose populations are declining. Although such actions are culturally challenging (e.g., in Japan and Australia where immigration policies are strict but whose currently high median age is an issue), they are nevertheless

---

11. Elderly dependency ratio (EDR) = (people aged 65 and over/people aged 15–64) × 100 and intends to compare the dependent part of society with the productive part. In the future, if retirement age increased to, say, 70, EDR = (people aged 70 and over/people aged 15–69). This new calculation reduces EDR since elders decrease and working age people increase.

critical. Actions here eventually allow national median ages to converge. To be successful, receiving nations must plan to increase the education and training of incoming workers.

11.4.3 Ongoing Efforts: Retirement Laws, Immigration, Retraining, and Incentives

As you might guess, retirement-related changes are unpopular and unexecutable, as illustrated by the French. In 2010, France's President Sarkozy raised France's retirement age from 60 to 62 to "save France's money-draining pension system" (MSNBCNews, 2010). His initiative sparked worker protests and cost him political clout. In 2012, new President François Hollande reversed Sarkozy's actions (Rowley, 2012). Hollande will pay additional pension costs by increasing the number of years people must work to receive a full pension (DiLorenzo, 2013). However, these changes do not go into effect until 2020 – no doubt long after his term is up. This case underscores the fact that political leaders must be courageous to look globally and into the future, and they must act immediately. It also says that sustainability goals must be consistent over the long term, regardless of who is in power. Politicians cannot "kick the can."

Another interesting initiative is in Japan, whose population is declining and whose fertility rate has been among the lowest for decades. In the 1990s, Japan instituted the Angel plan to support child rearing and encourage families to have more children.[12] In addition to introducing other plans that make having children attractive, Japan and other low-fertility rate countries are adapting in other ways: immigration, revised policies for retirement age and benefits, retraining older workers, and incentivizing companies to hire older workers.[13]

All these examples show that individual nations use suboptimal and conflicting solutions to problems that originate from the same source, but whose effects are opposite. Proposed actions in this area will create more consistency among nations and will move us all closer to the goals of sustainability.

This chapter promotes taking actions on the basis of a deeper appreciation for sustainability and for the effects of our current lifestyles. In all cases, actions alter the rules or structures that underlie relationships in the system diagram. Increasing energy costs, decreasing population by reducing birth rates, and modifying individuals' retirement plans engender complex economic and social issues that require debate, compromise, and allegiance to the same end goal. In all these areas, to achieve sustainability we must stop suboptimizing solutions and start collaborating as members of the same global community. The next chapter moves us away from these sticky political issues and into

---

12. Boling (1998). The Angel plan was underfunded and ineffective.
13. Bonnett (2009); for example, the "plus one proposal" encourages families to have one more child. See also Boling (1998) and Zoubanov (2000).

the realm of technology and policies that improve the environment, energy sources, and food and water. It proposes actions that will smooth our transition as we pursue sustainability.

REFERENCES

Associated Press, 2013, June 15. China announces 'tough' pollution reduction measures. CBCnews/World. Retrieved from <http://www.cbc.ca/news/world/china-announces-tough-pollution-reduction-measures-1.1312343>.

Bhaumik, S., 2007, January 10. Cash boost for tribal families. BBC News. Retrieved from <http://news.bbc.co.uk/2/hi/south_asia/6215497.stm>.

Boling, P., 1998. Family policy in Japan. Journal of Social Policy, vol. 27, no. 2, pp. 173-190. Retrieved from <http://homepages.wmich.edu/~plambert/boling.pdf>.

Bonnett, A., 2009, March 5. The plus one policy: Japan's rapidly falling population has sparked an anguished debate: Should the country open itself up. NewStatesman. Retrieved from <http://www.newstatesman.com/asia/2009/03/ja>.

Brown, L., Gardner, G., Halweil, B., 2000. Beyond Malthus: Nineteen Dimensions of the Population Challenge. Earthscan, London.

DiLorenzo, S., 2013, October 15. France's retirement reform: too little, too late? The World Post. Retrieved from <http://www.huffingtonpost.com/2013/10/15/france-retirement-reform_n_4100270.html>.

EIA, 1998. Impacts of the Kyoto Protocol on U.S. energy markets & economic activity. U.S. Energy Information Administration. Retrieved from <http://www.eia.gov/oiaf/kyoto/kyotorpt.html>.

Eurostat, 2013, October. Environmental tax statistics. European Commission Eurostat. Retrieved from <http://epp.eurostat.ec.europa.eu/statistics_explained/index.php/Environmental_tax_statistics>.

Generation Investment Management, 2012. Sustainable capitalism. Generation Investment Management, LLP., UK. Retrieved from <www.generationim.com/media/pdf-generation-sustainable-capitalism-v1.pdf>.

Global Post, 2011, July 2. India's "sex drive" solution to population explosion. Global Post. Retrieved from <http://www.globalpost.com/dispatches/globalpost-blogs/weird-wide-web/indias-sex-drive-solution-population-explosion>.

International Monetary Fund, 2013, October. World Economic Outlook Database. Retrieved from <http://www.imf.org/external/pubs/ft/weo/2013/02/weodata/weoselagr.aspx>.

Levinson, A., 2007, July. Taxes and the environment: what green taxes to European countries impose? Tax Policy Center, Brookings Institute. Retrieved from <http://www.taxpolicycenter.org/briefing-book/key-elements/environment/europe.cfm>.

Lim, L., 2008, April 14. China demographic crisis: too many boys, elderly. NPR. Retrieved from <http://www.npr.org/templates/story/story.php?storyId=89572563>.

Mongabay, 2008, December 7. World Statistics. Retrieved from <http://www.mongabay.com/igapo/world_statistics_by_area.htm>.

MSNBCNews, 2010, November 10. France raises retirement age despite protests. Europe on NBC News.com. Retrieved from <http://www.nbcnews.com/id/40103988/ns/world_news-europe/t/france-raises-retirement-age-despite-protests/#.UyDA-M-Ybq4>.

Park, M., 2013, December 28. China eases one-child policy, ends re-education through labor camps. CNN. Retrieved from <http://www.cnn.com/2013/12/28/world/asia/china-one-child-policy-official/index.html>.

Rowley, E., 2012, June 6. French president Francois Hollande cuts retirement age. The Telegraph. Retrieved from <http://www.telegraph.co.uk/finance/financialcrisis/9314666/French-president-Francois-Hollande-cuts-retirement-age.html>.

The Economist, 2008, July 10. The marathon's not over. The Economist. Retrieved from <http://www.economist.com/node/11708001>.

The World Bank, 2014. Fertility rate, total (births per woman). Retrieved from <http://data.worldbank.org/indicator/SP.DYN.TFRT.IN>.

Weiss, K., Mostaghim, R., 2012, July. Iran's birth control policy sent birthrate tumbling. Los Angeles Times. Retrieved from <http://www.latimes.com/news/nationworld/world/population/la-fg-population-iran-20120729-html, 0,4861001.htmlstory#axzz2uBuJUcmy>.

Zoubanov, A., October, 2000. Population ageing and population decline: government views and policies. New York: Department of Economic and Social Affairs, United Nations Secretariat. Retrieved from <http://www.un.org/esa/population/publications/popdecline/Zoubanov.pdf>.

Chapter 12

Pieces of the Puzzle Level III: Transition to the Future

Modern technology has introduced actions of such novel scale, objects and consequences that the framework of former ethics can no longer contain them. ...an object of an entirely new order – no less than the whole biosphere of the planet – has been added to what we must be responsible for because of our power over it.

– Hans Jonas (Jonas, 1984)

In the previous two chapters, changing our paradigms and altering the structure of the system that governs sustainability have necessarily focused attention on the future. Now, we wonder what we can do about the present as we move into the future; what actions can we take to ease our lives during the transition when proposed changes degrade our lifestyles and break our addiction to growth. This chapter attends to those concerns and completes the framework for our sustainability plan. Like placing the final pieces in the globe puzzle in Fig. 12.1, actions at this third level finally let us view the whole picture.

Third level actions promote investment in technology and innovative policies to alleviate today's stressful conditions. These stopgap measures allow us to live a little longer within Earth's carrying capacity and buy us time in our transition to a sustainable world. Because these actions are neither structural nor value related, they are only moderately effective but far less challenging.

FIGURE 12.1   Level III intervention.

K.L.Higgins: Economic Growth and Sustainability. http://dx.doi.org/10.1016/B978-0-12-802204-7.00012-8

In fact, many current efforts that attend to urgent energy, pollution, food, and water issues appear at this level. Considering that we are barely making a difference in our pursuit of sustainability, we can appreciate that actions at this level *cannot* be independent of actions at the other two levels, or we will fail.

12.1 ATTENDING TO THE PRESENT

Intervention areas 6, 7, and 8, as shown in Fig. 12.2, smooth the transition away from current lifestyles. Actions here can be implemented at national or local levels and require money, creativity, and global collaboration to stimulate innovation. They encourage technology solutions and policies that: (1) improve the *environment*; (2) increase efficiency and availability of *energy*; and (3) ensure adequate supplies of *food and water*. The greatest challenge at this level is to ensure that actions do not unintentionally cause future damage. Dashed lines in the system diagram trace their immediate influences.

The arrows of influence from these three areas eventually touch all three sustainability components; they decrease *damage to the environment* and its resultant *repair costs* and *weather extremes*; stabilize *the global economy* by reducing the *rate of energy consumption* and increasing *energy available*; increase *sustainable happiness and well-being* through a cleaner environment; and diminish *conflict, starvation, and disease* by supplementing *food and water available*.

12.2 LONG-TERM VERSUS SHORT-TERM CONFLICT RESOLUTION

Because we are dealing with immediate problems, technology initiatives and policies in these three areas are particularly susceptible to short-term thinking. Actions must consider whether their long-term effects will exacerbate the very problem they intend to remedy. Systems thinking labels this all-too-common conflict between present and future as "fixes that fail."

For example, consider what happens when we try to fix a single critical problem. In Fig. 12.2, increasing the amount of *food and water available* (positive action) intends to reduce *starvation and disease* (positive outcome). However, using traditional methods that involve toxic chemicals causes *damage to the environment* (negative outcome). More *food and water available* ultimately increases *population* (negative outcome). Over time, more people stress food and water supplies which raises *conflict, starvation, and disease* and diminishes *sustainable happiness and well-being* (negative outcomes). In other words, our narrow fix has failed.

In another area, superficial attempts to reduce *damage to the environment* such as recycling (positive action) decrease *repair costs*, improve the *global economy*, and increase *happiness and well-being* (positive outcomes). However, long term, these actions temporarily take the pressure off the need to stabilize the *global economy* and allow us to justify its unbounded growth. Economic growth spawns more *energy consumption* (negative outcome) which eventually increases *damage to the environment* (negative outcome). Again, the fix has failed.

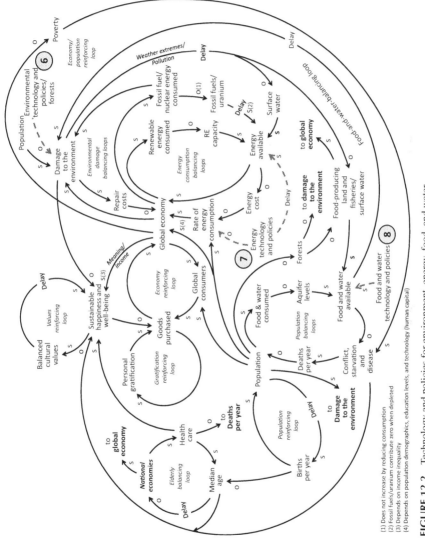

FIGURE 12.2 Technology and policies for environment, energy, food, and water.

(1) Does not increase by reducing consumption
(2) Fossil fuels/uranium contribute zero when depleted
(3) Depends on income inequality
(4) Depends on population demographics, education levels, and technology (human capital)

Both examples illustrate that the eventual effects of each well-intended action oppose original goals to reduce *damage to the environment*, increase *sustainable happiness and well-being*, stabilize the e*conomy*, and reduce *population*.

So what can we do? Do we avoid actions that alleviate current issues but undermine sustainability? With some caveats, the answer is no. If these actions prevent unmanageable or immediate devastation to economy, environment, or society, they must remain *as long as* other actions move us more forcefully toward sustainability. In other words, the *sum* of all actions must propel the system toward desired goals, but allow it to accomplish these goals with the least harmful repercussions.

In the above situations, future-oriented structural actions that offset undesired effects of short term fixes include reducing birth rates (intervention 4) and reducing sources pollution (intervention 6). Our plan must both temper current pressures that cry for immediate attention and ensure that today's policies do not degrade tomorrow's sustainability.[1]

12.3 AREA 6: ENVIRONMENTAL TECHNOLOGY AND POLICIES

"The environment." What an expansive term to describe the cradle of life! Earth's environment has many components: the air that we breathe; the climate that replenishes; water that refreshes and sustains; energy sources that fuel; and diverse plant and animal life that feeds, clothes, heals, and delights. Yet human activity threatens these necessities. To promote sustainability, actions in area 6 alleviate these threats. Notably, they: (1) reduce pollution; and (2) repair the environment and reforest.

12.3.1 Reduce Pollution

Even if world population stopped growing tomorrow, Earth's billions still produce extraordinary amounts of harmful pollution that degrades our quality of life and our survival. Topping the list is excess greenhouse gas (GHG) and the related issue of deforestation. Other pollutants from Chapter 6 are air pollution, municipal solid waste, radioactive waste, and industrial, agriculture, and human wastes that pollute land and water. The most industrialized parts of our society – we who revere economic growth – generate the bulk of these byproducts.

We begin with GHG because its effects are invasive and devastating. To lower GHG emissions, we first pinpoint its physical origins. Fossil fuel is the greatest contributor. Fig. 12.3 shows that two-thirds of global emissions come from energy consumption, nearly 90% of which involves fossil fuels.[2] Transportation and industry each account for over a fifth of this energy consumption; electricity and heat generation add another 37%. Therefore, to have the greatest effect on global GHG, we must focus on energy consumption.

---

1. See Forrester (1971) for discussion on short-term versus long-term policies.
2. BP (2014). Fossil fuels account for 87% of energy consumption in 2013.

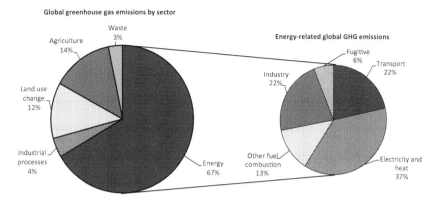

FIGURE 12.3   Global greenhouse gas emissions by sector (2008). *Source: Greenhouse gas data in this chart from Herzog (2009). "Fugitive Emissions" accounts for leakage from pressurized or other industrial equipment; "land use change" is primarily deforestation.*

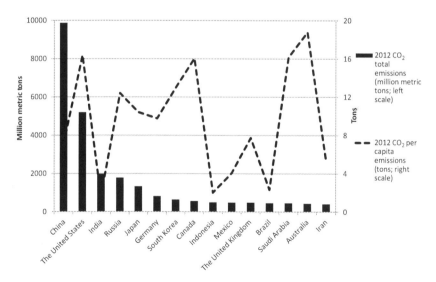

FIGURE 12.4   Highest CO_2 emitting countries (2012). *Source: CO_2 data from EDGAR (2014).*

Knowing the geographic origins of GHG in addition to their physical origins can help us tailor global actions. Fig. 12.4 shows that in 2012, China and the United States generated far more CO_2 than any other country. Furthermore, because the emerging economies of China and India are growing so rapidly and their populations are so huge, we expect their total GHG emissions to escalate over the next decade and their currently low per capita emissions to increase. Global efforts to decrease emissions must give priority to these countries.

Countries can also tailor efforts to manage their unique sources of CO_2. For example, in the United States, GHG emissions from transportation are twice the world average.[3] This distinction is not surprising since the United States has more motor vehicles per capita than any other country.[4] And China burns nearly as much coal as the rest of the world combined. Its economic growth depends on coal.

In addition to working with countries with huge *total* CO_2 emissions, actions should include countries with high *per capita* emissions. These countries must help citizens and industries decrease energy consumption. Fig. 12.4 indicates that emissions per person are elevated in the United States, Russia, Canada, Saudi Arabia, and Australia, but that, except for the United States, their total emissions are relatively low because their populations are small. Even so, individual energy use in these countries pales in comparison to Qatar, Kuwait, UAE, and Bahrain (not on the chart) whose per capita emissions are nearly double those of Australia (EDGAR, 2013). Actions in these countries should address their greatest source of emissions.

There are many ways to reduce fossil fuel emissions, all of which affect economies and people. The most obvious is to retard industrial processes and change personal habits, for, as Cleveland says, "rates of consumption are another name for rates of pollution" (Cleveland, 1993). Energy efficient appliances, hybrid automobiles, and clean manufacturing are part of the solution, but we must also wean ourselves from fossil fuel guzzlers. In the United States, driving fewer miles and reducing the number of gasoline-powered cars will decrease CO_2. For China and other coal-burning countries, technology to clean up fumes and transition to renewable energy or natural gas will reduce emissions. Many of these actions are doubly beneficial; they also diminish the rate at which we are depleting our limited caches of fossil fuels.

Fossil fuels are not the end of the GHG story. In 2011, agriculture, the second largest GHG emitter, accounted for 13% of total global emissions. Most of its emissions are nitrous oxide from fertilizers and wastes, and methane from cattle. In this case, countries with the largest agricultural emissions (e.g., China, Brazil, the United States, India, Indonesia, and Russia) should work with this industry. It is important to help farmers determine how to reduce emissions. Actions here include better crop management and conservation practices. At a global level, reducing food waste would decrease the amount of food produced. Globally, we use 198 million hectares (about the size of Mexico) to produce food which is simply thrown away.[5]

Finally, each year we convert forest land into food-producing land or other use (see Chapter 7). In the process we are nibbling away at a precious resource

---

3. EPA (2013a). Data from 2011; transportation contributed 14.3% of global total GHG and 28% in the United States.
4. The World Bank (2013). In 2010, in the United States, 797 people out of every thousand had motor vehicles versus China's 58 and India's 18. Second ranking Italy had 679 per thousand.
5. Agriculture data in this paragraph from Russell (2014).

that naturally removes CO_2 from the atmosphere. Thus, to reduce CO_2 emissions and increase CO_2 absorption, nations must manage their existing forests and increase reforesting efforts. For example, Brazil and Indonesia could augment current preservation, reforestation, and sustainable management efforts for their huge rainforests.

We are not yet finished. We must also decrease other harmful pollutants, that is, air pollution, municipal solid waste, industrial, agriculture and human wastes, and radioactive waste. Intervention 4 indirectly decreases pollution by reducing the number of people who generate it. This intervention merely slows down the rate at which pollution is increasing, but does not solve the problem. Intervention 8 described later in the chapter reduces agricultural waste by decreasing damage to land and water, but again this indirect effect is insufficient. Therefore, we must expand efforts to cut these pollutants at their source. Attending to root cause rather than remedying symptoms is the key to success in this area.

Because actions here are so broad and numerous, it would take another book to list them. Thus, we leave it to the reader to think about other initiatives that consider both immediate and secondary systemic effects. One important aspect of making these initiatives successful is to coordinate and increase ongoing independent and prolific efforts at all levels – individual, community, state, nation, and world. Such coordination can be achieved by setting rigorous global goals with specific milestones and ensuring that all levels know their roles.

12.3.2   Repair the Environment

The second part of this intervention involves environmental repair, including cleanup. This work is never ending, particularly in light of the many pockets of acute land, air, and water pollution. Efforts should become global imperatives, but can be accomplished locally. As in the case of GHG, actions must be tailored for a nation's specific environment. Each country must identify its greatest pollution issues. Ghana, for example, could implement policies that forbid burning of electronic or e-waste (see Chapter 6),[6] and invest in technology to safely dispose of e-waste and remove pollutants. Indonesia could apply technology and policies to clean up the Citarum and Ciliwung rivers, including finding alternative ways to dispose of waste from humans, agriculture, and livestock.

While these actions are somewhat unique to individual countries, they all fall under a global need to repair the environment. It is not enough to concentrate on local issues. Successful technologies and effective processes should be shared with all nations and communities. We must recognize that "answers will be found only by widening our worldview, changing our minds about the scale and dimension of what we face" (Cleveland, 1993).

---

6. E-waste includes electronic products that are no longer used, including old computers, televisions, copiers, and phones. E-waste that is not refurbished ends up in dumps or disposal sites. Many of these products release toxic chemicals when they are destroyed.

12.3.3 Ongoing Efforts: Reducing GHG, Reforesting, and Climate Engineering

Repairing the environment tops the ecological agenda of many nations. Like several other nations, the United States has had reasonable success with its initiatives. Its Clean Air Act of 1970 reduced six common pollutants by 72% even as the economy grew more than 200% (EPA, 2014). Primary reasons for this decline include: changes in fuel composition; emissions standards for transportation; clean air technology for plants and factories; reduced mercury waste from power plants; and devices that capture emissions from coal-fired power plants.

At the other end of the spectrum, Greenpeace International recently participated in the reforestation fests of Greenpop, a South African organization whose volunteers plant trees (Greenpeace, 2014). Even unlikely companies are getting involved; Paul Mitchell Salon Care Products advertise their support of Reforest'Action in women's magazines and alert their broad audience in the process.[7] These types of efforts are small but valuable they exemplify effective actions that can accomplish desired goals (e.g., reduce environmental damage) with minimal effects on other areas of the system (e.g., the economy).

On the global front, initiatives such as the 1997 Kyoto Protocol were first steps in recognizing the need to reduce GHG. After this protocol expired in 2012, some nations set their own GHG targets. Many are also "curbing deforestation and boosting renewable energy sources" or are "experimenting with cap-and-trade plans."

By 2020, Australia plans to reduce GHG by 5% below 2000 levels; Brazil will bring its emissions to 1994 levels and cut deforestation by 80%; Canada proposes a 17% drop in GHG from 2005 levels; Indonesia pledges 26% and the United States 17% below 2011 levels; the European Union proposes 20%, Japan 25%, and Russia 15% below 1990 levels. So that its economy can still grow, China promises to become 40% more energy efficient by 2015 instead of reducing emissions. South Africa is considering clean-energy options but will allow its emissions to increase until 2025 so that its coal-dependent economy can grow to support its rising population.[8]

The big news in June 2014 was that the US Environmental Protection Agency (EPA) plans to propose a 30% cut of CO_2 emission levels from existing US power plants by 2030. States will have the flexibility to decide how to achieve these reductions but are already nervous about the economic ramifications.[9]

Other types of initiatives are also numerous. The US EPA and the United Nations are collaborating to manage electronic waste (EPA, 2013b). Indonesia plans to clean its rivers, but is realizing that reducing the diverse causes of pollution will unintentionally lessen food supplies. They now recognize that "everything has to be integrated and coordinated" (Soebagjo, 2010).

---

7. See *Glamour* magazine.
8. National initiatives from NPR (2011).
9. Cooney (2014). Proposed reductions based on 2005 emission levels.

New ideas are hitting the market every other day. Harvard bioengineer David Edwards invented a novel food packaging product he calls "the WikiCell." With this edible membrane, he hopes to reduce the paper and plastics that people discard each day (Strauss, 2013). Scientists in the Netherlands found that spraying titanium dioxide on pavements removes significant amounts of nitrogen oxide (a greenhouse gas that creates smog) from the air.[10] UC Riverside students experimented with using titanium dioxide as an inexpensive roof coating. They calculated that spraying the compound on a million roofs could remove 21 tons of smog each day (Barboza, 2014).

An unconventional example of reducing GHG is to alter people's diets. Denmark, for instance, collects a tax on foods with high saturated fat, which reduces beef consumption and decreases emissions from livestock (fewer cattle are raised). This initiative also reduces health care costs, which benefits Denmark's economy and partly compensates for economic losses from a slightly weakened livestock industry. Similarly, although their actions do not reduce GHG, other countries tax unhealthy foods to reduce health care costs and improve their economies. Hungary taxes foods with high sugar, salt, and caffeine content; France taxes soft drinks; Finland taxes confectionaries.[11]

In other quarters, new technologies to reduce GHG emissions and slow global warming are building steam under the umbrella of "climate engineering." Once considered science fiction, ideas from the National Academy of Sciences, Jet Propulsion Laboratory, and Pacific Northwest National Laboratory are receiving serious consideration: ships that spew salt to block sunlight; mirrored satellites that deflect solar rays into space; and reverse power plants that suck carbon from the atmosphere (Halper, 2014). We do not know the long-term consequences of these concepts.

Unfortunately, some of these ideas are stuck in the bubble of our mental model; their tacit assumption is that economic growth – powered by precious fossil fuels that emit greenhouse gas – must continue. Such suboptimal efforts remedy symptoms rather than eliminate root causes. They reinforce our belief that technology will fix our malaise. If technology leaders would consider the broader context and concentrate instead on the sources of GHG, the world could make giant strides toward sustainability.

As we can tell from these examples, efforts to clean the environment are all over the map. The many policies are inconsistent, complex, and often contradictory. Some nations will strain their economies, whereas other nations grow their economies and freely emit GHG. Even when combined, localized efforts are insufficient to offset a predicted 30% rise of emissions in the next 20 years (Harvey and Macalister, 2014). But ... with all the efforts and interest from

---

10. Barboza (2013). The air-cleaning potential of putting titanium dioxide on surfaces (photocatalytic surfaces) has been known for several years. Recent experiments show its applicability in daily life.
11. Nations' taxes on food from OECD (2012).

governments, companies, media, and individuals, critical mass is building; soon GHG emissions and new solutions may creep into our everyday conversations and cultural norms.

Still, imagine how powerful it would be if we coordinated these diverse efforts! We could make tremendous progress if we viewed problems and solutions globally with a systems perspective – one that considers the combined influence of actions and mitigates unintended consequences. Programs to achieve coordination are ongoing. For example, the World Resources Institute's ACT 2015 (Agreement on Climate Transformation 2015) proposes a worldwide consortium to develop a framework for reducing carbon footprints (World Resources Institute, 2014). Such venues strengthen international cooperation and introduce novel ideas.

12.4 AREA 7: ENERGY TECHNOLOGY AND POLICIES

Energy technology and policies appear as a saving grace in our mental model. We want them to compensate for depletion of the carbon-based resources that feed our economies and our lifestyles. To supplement our energy needs as we drink more deeply from the energy well, we believe that new inventions and new ideas will transfer our energy dependence from polluting nonrenewable to nonpolluting renewable sources.[12] We expect that conservation and efficiency will prolong our access to carbon-based resources until technology kicks in. In this view, we believe we will never feel the pinch of energy shortages.

Chapter 6 demonstrated why this scenario is unrealistic. Energy consumption is escalating and there is no evidence that renewable or alternative sources can support our needs even in 25 years. Heinberg, for example, notes that wind power ranks high among alternative renewable energies, but its total potential "remains below the level needed to sustain the present scale of industrial society" (Heinberg, 2009).

Furthermore, there are drawbacks to betting our children's legacy on technology. One is that many renewable sources require immense infrastructures and have production and distribution challenges. Diamond suggests another weakness: "New technologies ...regularly create unanticipated new problems" (Diamond, 2005). A case in point is the automobile. It brought mobility and prosperity, yet it is a major source of CO_2.

But this bleak picture does not mean we should give up hope. It simply means that actions in this area must have two goals: (1) increase investment in clean renewable energy technologies to increase the probability of a breakthrough; and (2) stretch our nonrenewable resources as far into the future as possible. It also means that we must *simultaneously* intervene in other areas such as

---

12. Primary renewable sources are the Sun, Earth's heat, wind, water (rivers, lakes, tides, and oceans); nonrenewable sources are fossil fuels (coal, oil, and natural gas, biomass, and radioactive minerals). Unexplored sources include Earth's magnetic field, lightning, and sound.

energy cost structure (intervention 3) and births per year (intervention 4) which will reduce the rate of energy consumption.

12.4.1   Increase Investment in Renewable Energy Technologies

Bringing leading-edge technologies to maturity takes decades, brains, and a constant flow of cash. Recent progress in the ever-elusive nuclear fusion (Smith, 2014) and in harnessing solar lasers, oceans, and volcanoes gives us hope (McNicoll, 2013). Still, we must keep in mind that renewable energy technology is not *the* answer to our problems.

As noted earlier, multiple, diverse, and independent efforts may have the greatest payoff. We must open our viewfinders and consider far-out possibilities. Furthermore, the underlying science from these independent efforts must escape the confines of closely guarded intellectual property, yet encourage exploration and research. To accomplish this feat, governments, organizations, and businesses could increase investments, grants, and sponsorships for renewable energy solutions. They could widely disseminate research results, but allow patent protection for specific physical implementations that would encourage manufacturing.

12.4.2   Conserve Energy Resources

Strong policies for global conservation, energy usage, and pollution (beyond the 1997 Kyoto Protocol) would reduce consumption rates on a grand scale; such policies require unprecedented international collaboration. One idea for all this coordination is what sustainable development advocates Carley and Christie call "nested networks" (Carley and Christie, 2000). These networks link horizontally and vertically to integrate local, regional, national, and global efforts.

Using existing guidelines, nations, states, businesses, and communities could expand or tailor their own programs to conserve energy. Governments could build on current tax breaks or incentivize businesses and individuals to conserve; communities could expand recycling and conservation, and encourage new conservation norms. Utility companies could promote and reward conservation more than they are today.

In this latter case, it is important that individual consumers reap the rewards of their efforts. Requiring conservation without giving preferential treatment to conservers deters conservation. For instance, those who reduce their electricity usage should not only be charged less, they should also be rewarded with public recognition or refunds.

The last resort in this area is to constrain manufacturing or to ration fossil fuel and electricity. We hope these measures do not come to pass.

12.4.3   Ongoing Efforts: Green Energy and Energy Conservation

Many technology investments today support harnessing energy from clean renewable sources such as sun, water, and wind. Investment in solar plants, residential solar panels, and wind turbines described in Chapter 2 are a beginning. Furthermore, the growth of green energy companies and the establishment of the Renewable Energy Policy Network in 2005 signal progress.

Beyond these efforts, many countries have expanded their investments in renewables; global investment hit a record $279 billion in 2011. Since then, incentive policy uncertainties caused most countries to scale back, and lower costs for solar systems reduced investments; the $214 billion investment in 2013 was 23% lower than in 2011.[13] It will be important to prioritize renewable energy investments above profit or other expenditures.

In some nations, energy conservation flourishes. Many governments give tax deductions for insulation, energy efficient windows, and solar panels. China provides corporate tax incentives to enterprises that identify new ways to conserve energy (KPMG International, 2012). The US government has "partnered with manufacturing companies, representing over 1,400 plants, to improve energy efficiency by 25 percent over 10 years" (The White House, 2014). Even the US military is involved. In a cooperative agreement with industry, a Navy organization in California has developed a patent-pending process that converts CO_2 waste from its geothermal plant into jet and diesel fuels (Daily Independent, 2014).

Individuals, companies, and community initiatives are adding to the mix. British superstore Tesco operates its stores with wind power and rewards its management for meeting energy-reduction targets (Fortune, 2007). Solar panels in residential areas have also become more common. Private installation of photovoltaic cells in the United States hit an all-time high in the third quarter of 2013, surpassing Germany, the current world leader in solar capacity (Lundin, 2013). Small town recycling and highway cleanup, hybrid cars, efforts that protect local wildlife and water supplies are but a few examples. Utility companies incentivize energy and water efficient appliances; water companies offer rebates for buying rain barrels or soil moisture sensors and for removing grass lawns.

These endeavors, although encouraging, are too small and scattered to achieve planet-wide sustainability. They should be coordinated and expanded to all countries.

12.5   AREA 8: FOOD AND WATER TECHNOLOGY AND POLICIES

Day-to-day survival is the goal in this last area of intervention. Like the population aging dilemma, problems with *availability of food and water* are regional. Issues span the spectrum from starvation to obesity; some nations have abundant food and water while others cannot support their citizens' basic needs.

---

13. REN21 (2014). For the first time, in 2013, China invested more in renewable energy than all of Europe.

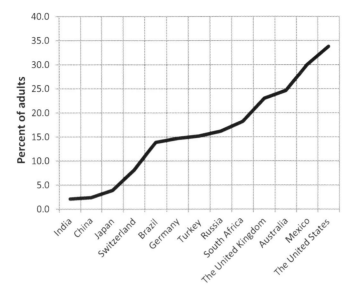

FIGURE 12.5   Obesity rates in selected countries (2009). *Source: Obesity data from OECD (2012) for 2009 or most recent available.*

A nation's economic health, culture, natural resources, population, and agricultural technology determine these diverse living conditions.

Obesity statistics in Fig. 12.5 paint a vivid picture of national differences. Because obesity is related to better living conditions and income, people in the United States were the most obese and people in India were the least obese of countries studied by the Organization for Economic Co-operation and Development. By the early 2020s, two-thirds of the people in many advanced nations will be obese.

In contrast to obesity, the United Nations estimates that about one in eight people in the world suffer from chronic hunger and cannot live an active life. Although sub-Saharan Africa has improved in the past decades, it still has the highest rate of undernourishment: about one person in five. Recent droughts and floods have significantly affected food supplies in these and other world regions.[14]

Given the stark contrast between these issues, it will be difficult to equalize needed interventions. Furthermore, as Chapter 9 notes, because this area increases availability of food and water, which *encourages* population growth, it contradicts intervention 4, which *discourages* it. The reason for this disparity is to alleviate suffering through a transition period when population is still growing and food and water supplies are inadequate. It is critical to reduce population growth in *all* countries *at the same time* that we attend to food and water.

---

14. FAO (2013). Uganda, for example, has a high population growth rate and a low agricultural productivity rate.

Actions to address this dilemma are not simply a matter of taking food and water from the prosperous and giving them to the poverty stricken. Responding to advertisements that ask for $1 a day to feed starving children does not solve the problem, and may exacerbate it by falsely encouraging population growth beyond a region's ability to sustain it. Instead, actions here must ensure that all nations have the knowledge and technology to maximize and supplement their natural resources.

But just how can we achieve this end? One answer lies with appropriate *food and water technology and policies*. Actions should: (1) increase food production; (2) conserve water; and (3) develop new fresh water sources. Note that actions in area 6 (discussed earlier) which mitigate *damage to the environment* also have a positive effect on food and water by reducing *weather extremes*.

12.5.1   Increase Food Production

Action here expands and globalizes technologies and policies that increase agriculture productivity without harming the environment. They freely transport excess food from one region to other needy regions, and they reduce food waste. Again, these actions must occur in *concert* with efforts to reduce population and thus diminish *demand* for food and water.

Technology could expand organic methods to increase productivity of existing land; decrease the toxicity of chemicals required to grow crops; and develop hardier and drought-resistant food types. Teaching better farming methods to countries whose practices are deficient would improve yield. Internationally conceived and locally applied policies that restore land that has lost topsoil or is oversalinized would increase productivity.

12.5.2   Conserve Water

As we noted in Chapter 7, we have dipped deeply into Ice Age aquifers – groundwater that has been stored water for millions of years. In the United States, for example, where aquifers furnish about half the drinking water and over 50 billion gallons a day for agriculture,[15] levels have dropped 25% in just 8 years (Konikow, 2013).

Furthermore, some regions are using more surface water than nature can replenish. Computer models predict that in North America alone "over the next 30 to 50 years ...changing current patterns caused by melting sea ice [will] increase average annual precipitation by 40 percent in the Northwest while decreasing it by 30 percent in the Southwest."[16]

These and other global forecasts suggest that drought will challenge many of the world's agricultural centers. Furthermore, the very process of growing

---

15. Keen (2012); see also Perlman (2014).
16. Park and Lurie (2014); Jacob Sewall at University of California-Santa Cruz conducted this research.

and transporting food damages the environment, which then decreases surface water and reduces food supplies. This network of cause and effect is definitely a sticky web. While water conservation is essential, it cannot be independent of reducing environmental damage.

Encouraging citizens to limit water consumption is a first step in conserving water. Yet, voluntary conservation is inadequate. Dr. Peter Gleick, codirector of the Pacific Institute, suggests that we must distinguish between voluntary and long-term improvements: "The first one is you let your lawn go brown. But the second one is that you replace that water-consuming lawn with drought-tolerant, native vegetation" (Park and Lurie, 2014).

In other words, local, sporadic, and panic-driven conservation is not the answer. Instead, we must alter how we think about water and not squander it before the situation becomes severe. Although such attitudes will change our lifestyles and our habits, they must be promoted worldwide. Increasing the price of water to reduce consumption and investing in research and infrastructure are essential; conservation may soon be mandatory.

Residential water is only a fraction of the problem. Because agriculture uses about two-thirds of our global water supply (Pearce, 2008) – as high as 80% in agricultural areas such as California (Pierson, 2014) – it is a prime target for water conservation. Again, technologies that support drought-resistant plants and make watering more efficient, and unpopular policies that penalize excessive use of water, will be crucial.

12.5.3   Develop New Water Sources

Though essential, water conservation alone does not mitigate our growing global water depletion. We must devise innovative ways to produce *more* fresh water and find energy efficient, cost-effective, and portable ways to desalinate water from the ocean (Rogers, 2014), to clean up rivers and build dams, and to install ecology-conscious pipelines that distribute water from precipitation-soaked regions to drought-plagued regions. We must use waste water more efficiently and capture precipitation before it flows into the ocean.

12.5.4   Ongoing Efforts: Sustainable Agriculture and Water Conservation

International agencies such as Greenpeace are already campaigning for sustainable agriculture to increase food production in the presence of drought.[17] We can build on their foundation by investing in technology research, education, and policies that promote crop rotation, cover crops, and new food variants.

---

17. Greenpeace promotes "ecological farming" to enable communities to produce enough food to feed themselves and to help cope with climate change. Retrieved from <http://www.greenpeace.org>.

Agricultural agencies and international organizations are also engaged; they seek to conserve agricultural water with measures such as using organic fertilizers to hold water and developing drought-resistant varieties of wheat, corn, and rice (Tirado and Cotter, 2010). They also recommend adjusting planting dates and crop variety, and improving erosion control (IPCC, 2007). On a smaller scale, researchers and various companies promote soil moisture sensors to manage water usage for growers (Peters, 2014) and recommend harvesting rainwater as well as storing and reusing water (IPCC, 2007).

In the conservation category, local governments in water-poor regions have restricted water usage (Keen, 2012; Park and Lurie, 2014). In California, for example, public utilities and the Governor urge citizens to cut water usage by 10–20% (Rogers, 2014; Park and Lurie, 2014). These stopgap measures fall short of what will be required for sustainability and tend to be abandoned when rain falls.

To intervene in the area of food and water, all nations must first appreciate the gravity of the situation. Policies and technologies – whether they apply to conservation or to production of water and food – should be coordinated nationally and internationally, regardless of whether particular regions have a current need. Without global attention, results will be disappointing.

The last three areas of intervention in this chapter smooth our transition to a sustainable world. They attend to near-term issues while ensuring that these actions do not have destructive long-term consequences. Furthermore, they augment actions that address issues du jour with stronger initiatives that move us closer to sustainability.

So far in the book, we have used a systems approach to visualize the complexity of sustainability, understand its interdependencies, and devise a strategy to achieve it. This chapter and the previous two described this strategy as an integrated and synergistic combination of eight areas of intervention with associated actions. Inherent in these actions are balance and integration – both of which are essential parts of the solution. The next chapter takes these lessons a step further and recommends how we can nurture sustainable stewardship as collectives and as individuals.

REFERENCES

Barboza, T., July 5, 2013. To clean the air, Dutch scientists invent pavement that eats smog. Los Angeles Times. Retrieved from <http://articles.latimes.com/2013/jul/05/science/la-sci-sn-smog-eating-street-20130705>.

Barboza, T., June 6, 2014. Smog-busting roof tiles could clean tons of pollution, study says. Science Now. Los Angeles Times. Retrieved from <http://www.latimes.com/science/sciencenow/la-sci-sn-smog-busting-roof-tiles-20140605-story.html>.

BP, June, 2014. BP statistical review of world energy 2014. Retrieved from <http://www.bp.com/en/global/corporate/about-bp/energy-economics/statistical-review-of-world-energy.html>.

Carley, M., Christie, I., 2000. Managing Sustainable Development, second ed. Earthscan, London.

Cleveland, H., 1993. Birth of a New World. Jossey-Bass, San Francisco.

Cooney, P., June 3, 2014. EPA to propose 30 percent cut in carbon dioxide emissions from existing U.S. power plants. Reuters. Retrieved from <http://www.huffingtonpost.com/2014/06/01/epa-carbon-emissions_n_5428561.html>.

Daily Independent, June 12, 2014. NAWCWD leads way for alternate energy. Daily Independent 88 (116). Retrieved from <http://www.ridgecrestca.com/article/20140612/News/140619875>.

Diamond, J., 2005. Collapse: How Societies Choose to Fail or Succeed. Penguin, New York.

EDGAR, 2013. CO_2 time series 1990-2011 per capita for world countries. Emission Database for Global Atmospheric Research. Retrieved from <http://edgar.jrc.ec.europa.eu/overview.php?v=CO2ts_pc1990-2011>.

EDGAR, 2014. CO_2 time series 1990-2012 per region/country and CO_2 time series 1990-2012 per capita for world countries. Emission Database for Global Atmospheric Research. Retrieved from <http://edgar.jrc.ec.europa.eu/index.php>.

EPA, September 9, 2013a. Climate change: sources of greenhouse gas emissions. United States Environmental Protection Agency. Retrieved from <http://www.epa.gov/climatechange/ghgemissions/sources.html>.

EPA, December 16, 2013b. Cleaning up electronic waste (e-waste). United States Environmental Protection Agency. Retrieved from <http://www.epa.gov/international/toxics/ewaste/index.html>.

EPA, February 6, 2014. Progress cleaning the air and improving people's health. United States Environmental Protection Agency. Retrieved from <http://www.epa.gov/air/caa/progress.html>.

FAO, 2013. The state of food insecurity in the world: the multiple dimensions of food security. Executive Summary. Food and Agriculture Organization of the United Nations. Retrieved from <http://www.fao.org/docrep/018/i3458e/i3458e.pdf>.

Forrester, J., 1971. Counterintuitive behavior of social systems. Technology Review. Cambridge: Alumni Association of the Massachusetts Institute of Technology Updated March 1995. Retrieved from <http://clexchange.org/ftp/documents/system-dynamics/SD1993-01CounterintuitiveBe.pdf>.

Fortune, 2007. 10 green giants. CNN Money. Retrieved from <http://money.cnn.com/galleries/2007/fortune/0703/gallery.green_giants.fortune/4.html>.

Greenpeace, June 4, 2014. Africa: Greenpeace partners with Greenpop at Platbos reforest fest. Greenpeace International. Retrieved from <http://allafrica.com/stories/201406041394.html>.

Halper, E., March 4, 2014. Climate control efforts gather steam. Los Angeles Times.

Harvey, F., Macalister, T., January 15, 2014. BP study predicts greenhouse emissions will rise by almost a third in 20 years. The Guardian. Retrieved from <http://www.theguardian.com/business/2014/jan/15/bp-predicts-greenhouse-emissions-rise-third>.

Heinberg, R., 2009. Searching for a miracle: net energy limits & the fate of industrial society. International Forum on Globalization and the Post Carbon Institute. Retrieved from <http://www.postcarbon.org/new-site-files/Reports/Searching_for_a_Miracle_web10nov09.pdf>.

Herzog, T., July, 2009. World greenhouse gas emissions in 2005. World Resources Institute. Retrieved from <http://www.wri.org/publication/world-greenhouse-gas-emissions-2005>.

IPCC, November, 2007. Climate change 2007 summary for policy makers and working group III: mitigation of climate change. Intergovernmental Panel on Climate Change. Retrieved from <http://www.ipcc.ch/pdf/assessment-report/ar4_syr_spm.pdf> and <http://www.ipcc.ch/publications_and_data/ar4/wg3/en/ch1s1-3.html>.

Jonas, H., 1984. The Imperative of Responsibility: In Search of an Ethics for the Technological Age. University of Chicago Press, Chicago.

Keen, J., July 19, 2012. Midwest drought and heat increase water supply worries. USA Today. Retrieved from <http://usatoday30.usatoday.com/weather/drought/story/2012-07-19/weather-drought-dry-conditions/56343680/1>.

Konikow, L., 2013. Groundwater depletion in the United States (1900–2008). Scientific Investigations Report 2013-5079. US Geological Survey. Retrieved from <http://pubs.usgs.gov/sir/2013/5079/SIR2013-5079.pdf>.

KPMG International, June, 2012. Taxes and incentives for renewable energy. Retrieved from <http://www.kpmg.com/Global/en/IssuesAndInsights/ArticlesPublications/Documents/taxes-incentives-renewable-energy-2012.pdf>.

Lundin, B., 2013. 2013 to go down in solar history. FierceEnergy. Retrieved from <http://www.fierceenergy.com/story/2013-go-down-solar-history/2013-12-10>.

McNicoll, A., November 13, 2013. Solar lasers, ocean power and volcanoes: unusual energy sources of the future. CNN. Retrieved from <http://www.cnn.com/2013/11/13/tech/innovation/solar-lasers-ocean-power-energy/index.html>.

NPR, December 9, 2011. What countries are doing to tackle climate change. National Public Radio. Retrieved from <http://www.npr.org/2011/12/07/143302823/what-countries-are-doing-to-tackle-climate-change>.

OECD, 2012. Obesity update 2012. Retrieved from <http://www.oecd.org/health/49716427.pdf>.

Park, A., Lurie, J., February 10, 2014. California's drought could be the worst in 500 years. Mother Jones. Retrieved from <http://www.motherjones.com/environment/2014/02/california-drought-matters-more-just-california>.

Pearce, F., June 19, 2008, . Virtual water. Forbes. Retrieved from <http://www.forbes.com/2008/06/19/water-food-trade-tech-water08-cx_fp_0619virtual.html>.

Perlman, H., February 24, 2014. Groundwater depletion. The USGS Water Science School, US Department of the Interior/US Geological Survey. Retrieved from <http://water.usgs.gov/edu/gwdepletion.html>.

Peters, R., 2014. Practical use of soil moisture sensors for irrigation scheduling. Irrigation in the Pacific Northwest. Washington State University Extension, Oregon State University Extension, University of Idaho Extension. Retrieved from <http://irrigation.wsu.edu/>.

Pierson, D., February 4, 2014. Drought leaves dark cloud over California ranchers, growers. Los Angeles Times. Retrieved from <http://articles.latimes.com/2014/feb/04/business/la-fi-ranchers-drought-20140205>.

REN21, 2014. Renewables 2014 global status report. Renewable Energy Policy Network for the 21st Century. Retrieved from <http://www.ren21.net/Portals/0/documents/Resources/GSR/2014/GSR2014_full%20report_low%20res.pdf>.

Rogers, P., January 28, 2014. California drought: 17 communities could run out of water within 60 to 120 days, state says. Retrieved from <http://www.insidebayarea.com/science/ci_25013388/california-drought-17-communities-could-run-out-water?IADID=Search-www.insidebayarea.com-www.insidebayarea.com%20>.

Russell, S., May 29, 2014. Everything you need to know about agricultural emissions. World Resources Institute. Retrieved from <http://www.wri.org/blog/2014/05/everything-you-need-know-about-agricultural-emissions>.

Smith, M., March 4, 2014. Laser bombardment yields energy milestone. CNN. Retrieved from <http://www.cnn.com/2014/02/12/tech/innovation/energy-fusion/index.html>.

Soebagjo, N., February-March, 2010. Cleaning up Citarum River. Asiaviews. Retrieved from <http://en.citarum.org/node/207>.

Strauss, M., May 31, 2013. Would you eat something wrapped in a WikiCell? Smithsonian.com. Retrieved from <http://www.smithsonianmag.com/arts-culture/would-you-eat-something-wrapped-in-a-wikicell-75866739/>.

The White House, 2014. Energy, climate change and our environment. Retrieved from <http://www.whitehouse.gov/energy/securing-american-energy>.

The World Bank, 2013. Motor vehicles (per 1,000 people). Retrieved from <http://data.worldbank.org/indicator/IS.VEH.NVEH.P3>.

Tirado, R., Cotter, J., April, 2010. Ecological farming: drought-resistant agriculture. Greenpeace Research Laboratories, University of Exeter, UK. GRL-TN 02/2010. Retrieved from <http://www.greenpeace.org/international/Global/international/publications/agriculture/2010/Drought_Resistant_Agriculture.pdf>.

World Resources Institute, 2014. ACT 2015. Retrieved from <http://www.wri.org/our-work/project/act-2015>.

Chapter 13

From Bud to Blossom: Nurturing Sustainable Stewardship

And then the day came when the risk to remain tight in a bud was more painful than the risk it took to blossom.

– Anaïs Nin[1]

You and I are each corners of the social fabric, and as such, our practice must include the social sphere and its workings within and around us.

– Michael Stone (Stone, 2009)

There is something about spring flowers that rejuvenates. Whether we are gardeners who plant seeds in cultivated ground and pull pesky weeds, or admirers of nature's wonders, we cannot help but be impressed as the plants mature and blossom. Like the bud waiting to open in Fig. 13.1, our proposed plan for sustainability is not yet mature. This chapter adds definition to make its implementation feasible and to create the full-blown blossom of sustainable stewardship.

FIGURE 13.1 Rose bud and blossom. *Source: Reproduced with permission of the artist, Sherri Scofield.*

1. This poem is attributed to Anaïs Nin, a popular French-Cuban author in the 1970s.

K.L. Higgins: Economic Growth and Sustainability. http://dx.doi.org/10.1016/B978-0-12-802204-7.00013-X

13.1   FROM ANXIETY TO HOPE

Thus far in the book we have immersed ourselves in bleak statistics and uncomfortable anecdotes of today's trends which drive us away from sustainability. Even considering possible interventions, I personally felt overwhelmed by the seriousness of the situation. Perhaps you have similar feelings.

I was also disheartened that individual acts, when added together, can spawn such a tragic social dilemma. At present, most human activity favors today over tomorrow and promotes self-interest over community-interest. These biases are so strong that, unless recast, they will eventually lead to an overshoot-and-collapse scenario (see Chapter 1) where sustainability is out of our hands. They paint a dark and depressing picture of our future world and make us question whether human nature and sustainability are mutually exclusive. Indeed, we could sigh in resignation as we bury the fears for our children into the deep dirt of excuses.

But let us not go there. This chapter intends to bring hope. It encourages us to strive for something far more significant than anything else we may ever achieve. Hope rests with our human ability to solve problems, on the progress we have already made, and on the millennial generation whose outlook is more global. To turn hope into reality, we will polish our plan and rely on new insights about the relationships that govern sustainability. But before we do, let us review where we have been.

13.2   FROM MENTAL MODEL TO INTEGRATED SYSTEM

This book initially compared our predominant mental model to a big plastic bubble that restricts our behaviors and guards us from reality's assault. The systems interpretation of this model resembled a giant ant that munches on consumers, excretes waste, and seeks instant gratification. The metaphoric ant represents a growing world imbalance in which economy overshadows environment and hurts society, and increasing population pushes sustainability out of reach. By identifying current trends, the book then revealed why this influential mental model is flawed: economic growth cannot continue forever, technology is not a magic potion for dwindling resources, and population and pollution profoundly affect us.

13.2.1   Economic Roots Are Deep

Our partiality toward the economy has deep roots. It began in earnest in the 1700s when Scottish Economist and "father of capitalism" Adam Smith introduced his "invisible hand" of the market (Smith, 1937). Smith proposed that when we each endeavor to achieve our individual goals in an economy, the end result is good for the collective. Indeed, over the past several centuries we have grown used to buying and selling whatever we please; economic growth has given many of us a better quality of life. However, this individual freedom has also borne massive issues that Smith could not have imagined.

By downplaying society and environment in favor of economic growth and without the ethical moorings of the 1700s, we are, as ecologist Garrett Hardin says "locked into a system of 'fouling our own nest'" (Hardin, 1968).

But we have had help. Human institutions that encircle every part of our daily lives promote and encourage economic growth. From religions to political philosophies and from "hierarchical control of basic resources" to the "influence of power and money" (Gowdy, 2014), we are surrounded by cultural and structural reminders that economic growth is good, says John Gowdy, Rensselaer economics professor. These structural shapers of beliefs and behaviors blind us to reality.

13.2.2   Deriving Relationships from Current Trends

Moving from the cocoon of our mental model, the book blended today's economic, environmental, and societal trends into an integrated system of relationships to show that our beliefs deviate from what is actually happening today. This system combines both short-term and long-term effects of our actions. It demonstrates how perturbing one part (e.g., economic or population growth) disturbs other parts (e.g., energy consumption, waste generation, or environmental damage). Fig. 13.2 symbolically illustrates the difference between our mental model and the integrated system: The relationships among the three circles of sustainability must change.

Using size to represent relative importance, our mental model on the left emphasizes economy; tiny overlaps among the circles indicate weak connection. By focusing on economic growth and material consumption, we discount the relevance of social issues such as population growth and the criticality of environmental issues such as pollution and climate change. To represent the balance required for sustainability, the system on the right equalizes economy, society, and environment; their greater overlap reflects their interdependence.

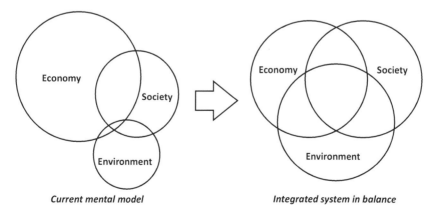

Current mental model   **Integrated system in balance**

FIGURE 13.2   Transition from mental model to integrated system. *Source: Figure adapted from overlapping circles of sustainable development defined at The World Conservation Congress in Bangkok, Thailand, November 17–25, 2004; IUCN (2004). Reproduced with permission of the International Union for Conservation of Nature.*

13.2.3   A Three-Level Framework for Intervention

From this integrated concept of sustainability, we developed a system diagram and located eight areas where intervention would be most effective. These areas framed a synergistic plan that attends evenly to economy, environment, and society. To differentiate their effects on sustainability, the areas were separated into three levels. The first lays the mental foundation for sustainability and shifts beliefs, cultures, and the predominant view of the world. The second advocates systemic structural changes that involve energy consumption, birth rate, and median age. Together, these actions reduce damage to the environment, decrease population, stabilize economic growth, and contribute to sustainable happiness and well-being. The third level bridges present and future by concentrating on urgent issues regarding energy, food, water, and environmental damage.

This framework fittingly emphasizes integration and balance, and delineates the responsibilities of collectives. However, it only provides the *whats* of achieving sustainability; it has not yet addressed *how* to coordinate the efforts of collectives or *how* to engage individual self-motivated acts. Incorporating these important aspects will move us from the nascent bud of a plan to the fleshed-out blossom of sustainable stewardship.

13.3   FROM INTEGRATED SYSTEM TO SUSTAINABLE STEWARDSHIP

To visualize the progression from bud to blossom, from the beginnings of a plan to a mature design, consider Fig. 13.3. Imagine that the simple circle triad on the left is the bud of a plan that, in time, grows into the lovely blossom on the

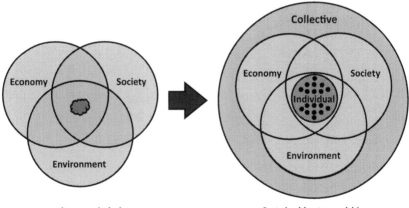

Integrated system in balance   *Sustainable stewardship*

FIGURE 13.3   Transition to sustainable stewardship. *Source: Figure adapted from overlapping circles of sustainable development defined at The World Conservation Congress in Bangkok, Thailand, November 17–25, 2004; IUCN (2004). Reproduced with permission of the International Union for Conservation of Nature.*

right. In the center of the blossom, individuals are the little pollen-producing stamens that create the seeds of future life. The surrounding outer green sepals define the collective and protect the inner pink petals of sustainability's primary concerns. Metaphorically, this transition brings ultimate balance to our system: Future is as important as present, and individual acts complement efforts of the collective. It incorporates the concept of sustainable stewardship in which we constantly renew our commitment to protect and be responsible for Earth.

Keep in mind that even with these additions, our plan is neither comprehensive nor a panacea. It is but a beginning – a way to approach the systemic problems posed by sustainability. It illustrates how the areas where actions will be most effective emerge from a systems perspective. Actions in this plan were easy to write down – the musts, shoulds, and oughts – but their implementation is daunting. Words on paper are inert; it is people who bring them to life. Sustainability requires the concerted efforts of many collectives and many individuals. Nations, institutions, and global citizens are all agents of sustainable stewardship.

13.3.1 The Collective

Let us first consider the collective component of sustainable stewardship. Our plan to achieve sustainability relies on the coordinated work of big groups – institutions, agencies, governments, and businesses. Evidence shows that such an approach is possible. Although they are suboptimal and often contradictory, national and community attempts to solve sustainability issues have had scattered success and demonstrate that we can move from sporadic collaboration toward a robust form of global governance.

13.3.1.1 Global Collaboration and Global Governance

Over the past seven decades, economic, social, and technological processes have made "the world more interconnected and interdependent."[2] With that evolution, pockets of society have joined forces to tackle the ever-escalating, ever-broadening world issues. Beginning in the mid-1940s, intergovernmental organizations such as the United Nations, The World Bank, and the International Monetary Fund increased global cooperation on common problems (Steger, 2013).

Countless informal collaborations and formal groups such as Greenpeace and the World Health Organization now manage issues of mutual interest, including those relating to environment and human health. More recently, the traditional G8 world forum (the eight wealthiest economies) expanded to G20 (twenty countries including emerging economies) to address the 2008 global financial crisis. Thus, many world powers recognize the serious and interdependent nature of global issues and are willing to put differences aside. The obvious question is: How can we make their collaboration more effective?

---

2. UCLA Geography Professor Laurence Smith (2010) provided this definition of "globalization."

The idea of "governance" instead of "government" offers a promising template for collective action and a way to get global "buy-in" of a sustainability plan. It moves us to take ownership of the issues and pushes us from a mindset of "they" to "we." Over the last two decades, this concept has taken root both in academic studies of human behavior and in practice, as exemplified by the boom in international nongovernmental organizations (NGOs). Governance, which "includes any system that humans use to make and enforce collective decisions," encompasses both formal rules and informal cultural values (Seyle and King, 2014). It supplants our narrow concept of nationalized, localized, and formalized governments that enforce laws and policies. Global governance is our promissory note to future generations; it can reflect our collective views about who we are, what we believe in, and what we wish to become (Cullinan, 2014).

Escaping the restrictions and self-interests of national, local, and community governments requires an international body dedicated to protecting Earth and to ensuring human well-being. This body would assimilate and expand today's attention on sustainability. Conceivably it could update and enlarge the United Nation's 1945 charter to maintain peace, international security, and respect for human rights. Its constitution would incorporate laws and guidelines for sustainability and for nations' rights with respect to common resources. Its activities would augment and coordinate current groups that address world issues.

13.3.1.2 Movement Toward Global Governance for Sustainability

Accomplishing shared governance at a global level is extremely difficult. It has never been done on the scale required for human survival. South African environmental attorney Cormac Cullinan shares several reasons why global governance is a challenge. Foremost among these is the need for a a cultural paradigm paradigm shift. Cullinan suggests that we must move from our "materialistic worldview" to what he calls "Earth jurisprudence" which dictates a deeper "understanding of the universe and our place within it" (Cullinan, 2014).

This need for a paradigm shift should sound familiar by now. It is the first and most essential area of intervention in our sustainability plan. Short of a worldwide environmental, ecological, or social disaster that unites us or a forceful takeover by a single entity that oppresses us, it will take widespread efforts in all parts of our society to accomplish this transformation of beliefs.

By definition, global governance must represent the views of all stakeholders, facilitate agreement, and negotiate resolutions. Unlike the United Nations, its membership cannot be restricted to a few nations and cannot be homogeneous. National governments (including emerging and undeveloped economies), international institutions (with various sustainability interests), religious organizations, businesses (from different industries such as energy, farming, finance, and manufacturing), media magnates, and spokespersons from our younger generations can all contribute a different perspective and must count among its members. To achieve the cultural transition and embed the values that will make global governance attainable, this group or groups must create a shared vision that quells differences and they must assume the power to reward and penalize.

At a national level, some governments have already begun to transform their cultural values. Bhutan and Bolivia, for example, are "reconceptualizing their governance systems in order to steer their country toward ecological sustainability." Ecuador has revised its constitution and its laws to recognize the legal rights of nature (Cullinan, 2014).

Other groups and venues, including the World Future Council and the 2012 U.N. Conference on Sustainable Development held in Rio de Janiero (Rio +20), acknowledge "intergenerational" concerns. Unfortunately, an initiative to appoint an "independent advocate for the welfare of future generations" was cut from the Rio +20 agenda. Although our youth have no formal voice yet, it is encouraging to find that "young people are being inspired in droves to take action, with many increasingly making it the defining battle of their generation" (Ebel and Rinke, 2014).

Other initiatives are luring businesses into the sustainability fold. With nearly 300 international member companies, the Business for Responsibility group has worked 20 years to shift business practices toward long-term value creation instead of short-term profit-taking. It is dedicated to driving "sustainability into corporate DNA" with a three-pronged approach of raising awareness, catalyzing change, and building systemic solutions (BSR, 2012).

Mainstream publications such as the *Harvard Business Review* urge businesses to collaborate on sustainability issues. One article identifies four models of "sustainable collaboration" based on work with international companies. It recommends that businesses start with a "founding circle of participants that share a common motivation and have mutual trust" (Nidumolu et al., 2014).

From these initiatives, we see progress in shaping what could become a form of global governance. Yet we wonder what the ultimate global governance might look like. A benevolent dictatorship would be too narrow and unrealistic, so let us expand our ideas and conjure up a scenario using a favorite science fiction series. In *Star Trek*, 150 planetary governments joined forces for "mutual trade, exploratory, scientific, cultural, diplomatic, and defensive endeavors." The Federation, as they called it, set guidelines for behavior; the greatest of these guidelines was the *prime directive* which prohibited members from "interfering in the normal development of any society" (Okuda et al., 1994). A pipedream? Maybe. But let us have great aspirations.

13.3.1.3   Potential Sources of Influence

In addition to collaborative enterprises that have already achieved success with global issues, the potential to influence the world's citizens is also growing; such influence will be essential for shifting paradigms and galvanizing commitment to sustainability. Among the many possibilities, five are most notable: (1) transnational corporations, (2) media, (3) formal education programs,

(4) the millennial generation, and (5) grass-roots organizations for sustainability. A global governance body could engage these influences.

1. *Transnational corporations:* Economic power has steadily migrated out of nations and into huge transnational corporations (TNCs). In 2011, TNCs accounted for about one-third of world output, three-fourths of world trade, and four-fifths of international investment (McGrew, 2011). Their revenues exceed the gross domestic product (GDP) of many nations; for example, in 2011 WalMart revenues surpassed the GDP of Saudi Arabia; Royal Dutch Shell revenues were greater than Poland's GDP.[3] The huge global reach and economic dominance of these organizations enhance their powers of persuasion. The challenge here is to cultivate their support before their assertive voices can stalemate sustainability.

2. *Media:* Another corner of influence comes not from nations nor TNCs: It is the culture-shaping power of the media. Like mist in a tropical forest, media saturates our consciousness – and our unconscious thoughts. The invasiveness of electronic media, exemplified by the Internet, television, and cell phones, and the pervasiveness of physical media, such as billboards, advertisements, and movies, make it difficult to escape the clutches of whatever message is transmitted. By showcasing preferences and lifestyles, such communication transmits cultural values. These values, or "cultural goods" as former US Vice President Al Gore calls them, "carry the cultural DNA of that country" (Gore, 2013).

   Media permeates us from the inside out. It conveys values without force. It changes perceptions. If used to communicate imperatives for future survival and drawbacks of current lifestyles, it can be sustainability's greatest ally. However, because it is so powerful, its downside must be managed. Not only can media give unrealistic impressions, including the glories of consumption, it is also managed by a tight group who controls its imagery. In 2006, eight conglomerates dominated the communications industry and "accounted for more than two-thirds of the US \$250–275 billion in annual worldwide revenues."[4] The power of these few can fall on either side of the tug-of-war between present and future, between today's economic growth and personal gratification and tomorrow's sustainability. Thus, their understanding and commitment to sustainability is essential.

3. *Formal education programs:* From preschool to graduate school, formal courses and educational programs profoundly influence children and young adults as their belief systems solidify. The influence of these programs on

---

3. Steger (2013). In 2011 WalMart revenues (\$446,950M) and Royal Dutch Shell revenues (\$470,171M); 2011 GDP for Saudi Arabia (\$434,666 M) and GDP for Poland (\$469,440 M); Steger refers to a 2011 study which found that 147 "super-connected corporations" control "40% of the total wealth" of over 43,000 large TNCs.

4. Steger (2013); the eight conglomerates were Yahoo, Google, AOL/Time Warner, Microsoft, Viacom, General Electric, Disney, and News Corporation.

sustainability mounts with each passing year. For example, in the United States, Virginia Beach City Public Schools involves their 70,000 students and 15,000 employees in sustainability practices and educating the public.[5] The College of Engineering at the California Polytechnic Institute, San Luis Obispo introduced a new program which focuses on sustainable energy generation and efficiency.[6] Grade school education that causes children to ask their parents why they let the water run when they brush their teeth, or urges them to pick up garbage, is not only creating awareness, it is shaping advocates.

4. *Millennial generation:* The next influence lives in our midst today; children and young adults will have immense sway in the near future. Consider the characteristics of the next-in-line leaders from the millennial generation.[7] We can feel heartened that they have a broader view of the world than previous generations. If made aware of the issues, their perspective promises open-minded and system-wide solutions. Judging from their frequent interactions on electronic devices and social media, exposure to other cultures and travel experience, they also have the potential to become excellent collaborators. As part of a global governance structure, leaders from the millennial generation – whose future will be disturbed the most by current trends – could counterbalance short-sighted or narrowly motivated plans.

Yet, even with all their passion and promise, these youth can only go so far with their advocacy for the future; they are, by definition, "resource-constrained and have little access to media or political power" (Ebel and Rinke, 2014). It is incumbent on those in power, on those with resources, and on older generations to engage and support them.

5. *Grass-roots organizations:* National governments that are separated by borders, regulations, politics, and insular economic concerns are hard-pressed to resolve global issues. Various grass-roots organizations recognize this dilemma between national and global issues; many are gathering voices to speak for the collective.

Today, in the absence of formal global governance, several influential individuals in key European nations have begun a campaign to unite world citizens into a powerful voting bloc. This "Simultaneous Policy (Simpol) campaign" encourages people to "use their votes in a new way to drive governments to solve problems like climate change, out-of-control financial markets, and social injustice."[8] The list of supporters grows each day.

---

5. See <http://www.vbschools.com/SustainableSchools/>.

6. See: A Wave of Sustainability, *Cal Poly College of Engineering Advantage,* Spring 2014.

7. The millennial generation includes those born between 1982 and 2003. It is the largest generation to date.

8. International Simultaneous Policy Organisation (2014). Among Simpol's supporters are thought leaders, such as Ken Wilber, who have long advocated a systems approach to problems.

Another example, the Solar Roadways project, is worth mentioning for two reasons.[9] First, it proposes to use solar panels instead of tar, concrete, or asphalt on roadways, driveways, and parking lots to generate solar energy and reduce greenhouse gases. A second more important reason in the context of this discussion is how this idea is promoted. Conceived in 2006 by two engineers, Julie and Scott Brusaw, the solar roadways invention is publicized and requests sponsorship over the Internet. Readers can buy a panel or receive hats, mugs, and shopping bags when they support the project. Such use of electronic media brings like-minded individuals together without regard for national or political orientation, and with the potential to launch the project globally. Other grass-roots ideas could benefit from this same approach.

13.3.2 The Individual

Because we have already witnessed pockets of collaboration among diverse and often competitive entities, and because there are many sources of influence, we can be optimistic that a properly implemented plan for sustainability can be successful. At its current rate of growth, support for sustainability may soon reach critical mass – the point at which sustainability is the norm rather than the exception and global governance is possible.

Yet, collaboration, group charters, and influence are not enough. This book initially asked whether it is possible to reconcile the dual nature of human beings. Can we create a balance between our present-oriented, self-focused natures and the need for future-looking, community-interest? The answer rests with individual behavior. We must learn how to tap the altruistic, communal side of human nature on a grand scale and add the role of the individual to our plan.

13.3.2.1 Individual, Self-Motivated Behaviors

In the metaphoric flower from Fig. 13.3, individual contributions lie in the sweet spot where economy, environment, and society overlap. These contributions are contained within a circle of the collective. Interpretation of this depiction is twofold. First, individuals can be members of groups or collectives that promote sustainability. Second, individuals whose actions are self-motivated can together create an informal and individually autonomous collective.

In the first case, our plan recommends actions and structures to engage collectives. In the second case, we know from experience that small, self-initiated individual contributions, when combined, can make a difference. When billions of separate and isolated acts unite, their power is greater than any formal structure. Consider for example the simple act of throwing a plastic bottle into a river or onto the street. If everyone in our community does the same thing and if no one cleans up after us, visualize what we will see in a year (recall Indonesia's

---

9. Indiegogo (2014). Other theoretical advantages keep roads ice-free, provide channels for power cables, and charge electric vehicles.

polluted river in Chapter 6). Now, suppose that instead of littering, every day we each pick up one piece of garbage that is not ours. The result: a clean river or a clean street in short order.

But the real dilemma is not what constructive behavior *could* look like when millions of separate acts are combined; instead, it is *how* to engage individuals in this constructive behavior. The solution and great challenge is to understand and engage basic human nature.

13.3.2.2   Understanding Human Nature

For centuries, understanding human nature has absorbed philosophers, psychologists, academics, business leaders, marketeers, and snake oil salesmen alike. How can I get someone to do something? Regardless of whether that something is for selfish or for altruistic reasons, this question underpins the idea of motivation. Answers to the question are particularly important if we expect to achieve the onerous goals of sustainability. Because one person's actions cannot begin to touch these goals, it is easy for individuals to say "not my problem" or "someone else will handle it" and go on about their business with the same mental model and the same behaviors that undermine sustainability. To alter individual behaviors, as our sustainability plan requires, we must understand what might prevent this change and what might promote it.

What prevents behavior change? Individuals refuse to change their behaviors relating to sustainability for many reasons. Monty Hempel, Environmental Studies Professor at the University of Redlands, has compiled a list of 21 factors that contribute to what he calls "Eco-Complacency and Disbelief" (Hempel, 2014). A small sampling of his factors, with my own adaptations, includes:

- psychological distance (not my problem)
- disinformation (man's effect on the environment is minimal)
- behavioral momentum (habits)
- lack of self-efficacy (one individual cannot make a difference)
- lack of place attachment (I can move away from the problem)
- complexity (there is no solution)

Because they reveal human nature, these factors (and Hempel's remaining factors) apply to *all* our proposed actions. Decision-makers should address each one as they implement sustainability policies and plans. For example, promoting awareness and education, and delineating specific personal consequences of individual actions could shatter psychological distance.

What promotes behavior change? In addition to offsetting obstructive individual responses, policymakers must also engage the aspects of human nature that stimulate change and elicit desired behavior. In this category, actions must be:

- inspiring (the vision of a possible future is compelling and meaningful)
- relevant (each of us knows how our personal lifestyle harms the Earth and how this damage affects us and our loved ones)

- fair (every person/every nation contributes to the goal, even when contribution means sacrifice)
- measurable (contributions are quantified to reinforce that is it worth the effort)
- incentivized (people are rewarded for changing their habits and lifestyles).

13.3.2.3 Individual Actions

Having identified ways to prevent destructive behaviors and promote constructive behaviors, we now move to the center of our metaphorical flower. Let us consider what each of us as an individual can do without the impetus, directions, or resources of formal collectives or global governance. Instead of combining so many acts of self-interest, we can unite solitary practical acts of concern for community and for the future.

When multiplied by the billions of people in the world, small acts and habit changes of one individual *matter*. What if every individual were to use one less gallon of water each day, use one less gallon of gas, or eat one less hamburger a week? Even multiplied by a million people, a gallon of water per day for a year translates into 5½ million pounds of potatoes[10] and a gallon of gasoline every week for a year translates into over half a million trips to deliver food or goods between New York and San Diego.[11] A quarter-pound burger per week for a year multiplied by a million people is about 25,700 cows that produce 5.6 million pounds of methane and drink 84.5 million gallons of water a year[12] – enough for 22,000 people in rural India or 3,600 Americans.[13] Still, we must keep in mind that, even for their benefits, these small acts have economic repercussions on those who sell cattle or deliver goods for their livelihoods. We can not have it all.

Vigorous local initiatives can spark change. We can take advantage of human nature here; when people identify with group goals, they make sacrifices out of loyalty to others in the group (Gardner and Stern, 1996). To see astonishing results, each of us can flush our toilets less; xeriscape our yards; give up a luxury; waste less food; take one less trip; set our thermostat one degree cooler in winter and one degree warmer in summer; recycle, clean up our trash; take family planning to heart; support those who lead community, national, and global sustainability initiatives; and nurture future leaders.

---

10. It takes 65 gallons of water to grow 1 pound of potatoes; see Pearce (2008).
11. Assumes 30 mpg and a distance of 2,834 miles.
12. One cow equates to about 2,024 quarter-pound burgers and produces 70–120 kg of methane per year (retrieved from <http://wiki.answers.com/> and <http://greenanswers.com/>). The average 1,100 pound (nonlactating) cow drinks 9 gallons of water a day or 3,285 gallons a year (retrieved from <http://www.extension.org/>. Americans averaged 58 gallons of water a year in 2013 (retrieved from <http://www.theatlantic.com/health/archive/2013/03/how-much-water-do-people-drink/273936/>.
13. United Nations (2014). In India the norm for rural water supply is 40 L (10.6 gallons) per person per day; for America (without outdoor irrigation, swimming pools, and leakages) it is 242 L (63.9 gallons) per person per day – with irrigation and pools, it is up to 650 L (172 gallons) per person per day.

We can make a difference by publically standing up for sustainability. Even unconventional approaches will work. For example, parents in one Wyoming town addressed their 10-year-old daughter in a letter to the editor explaining why they were voting for a particular bond issue. They told her (and the paper's readers) that "if everyone invests in your future, over time they will discover that you are well worth it. ...We do not want to put your future at risk" (Casper Star Tribune, 2014). Using whatever creative, novel or mundane way we can think of, we must move our thoughts and behaviors from ownership of individual pieces to stewardship of the whole.

Coordinated actions of collectives will move us closer to a balanced state of sustainability. Such coordination is difficult, but possible using the framework of global governance. However, collectives and formal programs are only part of the solution. As individuals, we must each know the problem and own it. We must ask: How am I making the problem worse? Do my actions consider community well-being and future consequences?

Thus, a blend of collective and individual participation brings us closer to the sustainable stewardship required to ensure that the needs of future generations can be met. As the old West African proverb says "it takes a village to raise a child" (Cowen-Fletcher, 1994). It also takes a village to achieve sustainability – a global village of Earth's stewards.

REFERENCES

BSR, 2012. BSR at 20: Accelerating progress. Business for Social Responsibility. Retrieved from <http://www.bsr.org/pdfs/reports/bsr-at-20-report.pdf>.

Casper Star Tribune, April 15, 2014. Letter to the Editor from Mom and Dad. Casper Star Tribune.

Cowen-Fletcher, J., 1994. It Takes a Village. Scholastic, Inc, New York.

Cullinan, C., 2014. Governing People as Members of the Earth Community. Governing for Sustainability: State of the World 2014. The Worldwatch Institute. Island Press, Washington, pp. 72–81.

Ebel, A., Rinke, T., 2014. Listening to the voices of young and future generations. Governing for Sustainability: State of the World 2014. The Worldwatch Institute. Island Press, Washington, pp. 82–90.

Gardner, G., Stern, P., 1996. Environmental Problems and Human Behavior. Allyn and Bacon, Boston.

Gore, A., 2013. The Future: Six Drivers of Global Change. Random House, New York.

Gowdy, J., 2014. Governance, Sustainability, and Evolution. Governing for Sustainability: State of the World 2014. The Worldwatch Institute. Island Press, Washington, pp. 31–40.

Hardin, G., 1968. The tragedy of the commons. Science 162, 1243–1248.

Hempel, M., 2014. Ecoliteracy: knowledge is not enough. Governing for Sustainability: State of the World 2014. The Worldwatch Institute. Island Press, Washington, pp. 41–52.

Indiegogo, 2014, May. Solar Roadways. Retrieved from <https://www.indiegogo.com/projects/solar-roadways>.

International Simultaneous Policy Organisation, 2014, May 27. Simpol's "new way to vote" builds support for global justice in EU Parliament. Retrieved from <http://www.simpol.org/fileadmin/user_upload/Articles/PR_Euro_2014_Results.pdf>.

IUCN, 2004. The IUCN Programme 2005–2008: Many Voices, One Earth. Adopted at The World Conservation Congress, Bangkok, Thailand 17–25 November 2004. Retrieved from <http://cmsdata.iucn.org/downloads/programme_english.pdf>.

McGrew, A., 2011. Globalization and global politics. In: Baylis, J., Smith, S., Owens, P. (Eds.), The Globalization of World Politics: An Introduction to International Relations. fifth ed. Oxford University Press, Oxford.

Nidumolu, R., Ellison, J., Whalen, J., Billman, E., 2014. April. The Collaboration Imperative. Harvard Business Review 92 (4), 76–84.

Okuda, M., Okuda, D., Mirek, D., 1994. The Star Trek Encyclopedia: A Reference Guide to the Future. Pocket Books, New York.

Pearce, F., June 19, 2008. Virtual water. Forbes. Retrieved from <http://www.forbes.com/2008/06/19/water-food-trade-tech-water08-cx_fp_0619virtual.html>.

Seyle, D., King, M., 2014. Understanding Governance. Governing for Sustainability: State of the World 2014. The Worldwatch Institute. Island Press, Washington, pp. 20–28.

Smith, A., 1937. An Inquiry into the Nature and Causes of the Wealth of Nations. Random House, New York.

Smith, L., 2010. The World in 2050: Four Forces Shaping Civilization's Northern Future. Dutton, New York.

Steger, M., 2013. Globalization: A Very Short Introduction, third ed. Oxford University Press, Oxford.

Stone, M., 2009. Yoga for a World out of Balance: Teachings on Ethics and Social Actions. Shambhala, Boston.

United Nations, 2014. Water and energy, volume 1. United Nations World Water Development Report 2014. United Nations Educational, Scientific and Cultural Organization, New York. Retrieved from <http://unesdoc.unesco.org/images/0022/002257/225741E.pdf>.

Chapter 14

The Global Commons and the Uncommon Globe: System Insights and Conclusions

When you go around the Earth in an hour and a half, you begin to recognize that ...from where you see it, the thing is a whole, and it's so beautiful.

– Astronaut Russell Schweikart[1]

The problem is not the management of the global commons. It is the management of human behavior in the commons.

– Harlan Cleveland (Cleveland, 1993)

Through the eyes of astronauts we know the uncommon beauty and absolute oneness of the tiny planet we call home. If you have never experienced their vantage point, it is hard to grasp that Earth's references for time and space – the passing of days and the sensation of standing still – are irrelevant. Although pictures of Earth taken from space are fantasy to those of us tethered here, these space pioneers are awed by the view (see Fig. 14.1). Even so, their mental shift is more impressive than the image they see, as astronaut Russell Schweikart recalls from his humbling 1968 space flight:[2]

"Somehow you recognize that you're a piece of this total life... And when you come back there's a difference...in that relationship between you and that planet, and you and all those other forms of life on that planet."

That Earth is a speck compared with the rest of the universe is inconceivable, but we do recognize one thing: Our globe is extraordinary. When we lower ourselves from the surrealism of space and onto reality's firm ground, we know that sustainability on Earth means that we, like the astronauts, must develop a new relationship with the resources we share with others – the "global commons," as they are called.

---

1. White (1987) from *No Frames, No Boundaries*, in *Earth's Answer: Explorations of Planetary Culture at the Lindesfarne Conferences*, West Stockbridge, MA: Lindesfarne/Harper & Row, 1977.
2. From *No Frames, No Boundaries*, film commentary as quoted in White (1987).

K.L. Higgins: Economic Growth and Sustainability. http://dx.doi.org/10.1016/B978-0-12-802204-7.00014-1

FIGURE 14.1 Earth from space. *Source: Earth's Eastern Hemisphere courtesy of NASA. Retrieved from <http://visibleearth.nasa.gov/view.php?id=57723>.*

This final chapter steps back from the minutiae of system diagrams and specific actions to view sustainability as though we were astronauts seeing Earth for the first time. Now that we have analyzed Earth's finite capacity and have the framework for a sustainability plan, we can use this telescopic perspective to appreciate the complexity of managing our global commons – especially in light of the global governance discussed in Chapter 13. Once we soak up the significance of this challenge, we then conclude the book with three insights gained from applying systems thinking to sustainability. We will find that these insights are already woven into our plan of intervention. This grand finale intends to tie up loose ends and smooth the potholes of the arduous road ahead.

14.1 THE GLOBAL COMMONS

The fundamental success of sustainability hinges on how we manage Earth's shared resources even as we address economic expansion, population growth, and environmental pollution. Although we have previously listed the most significant of these resources, described current trends that threaten their continued presence, and recommended a plan to reduce our consumption of them, we are left with questions of why they are so difficult to manage and why we have not done more to prevent their depletion. The answer rests with the very nature of human beings.

To explore the issues of managing our global commons, let us introduce the social dilemma called "tragedy of the commons." The concept of common resources, a "commons," is centuries old. It originated in medieval times when farmers did not own, but shared pastures to raise their livestock. Nearly 50 years ago, Garrett Hardin applied this idea to other areas of human life including population, food, pollution, energy, sound waves, national parks, and airspace (Hardin, 1968). In what he called "tragedy of the commons," Hardin compared

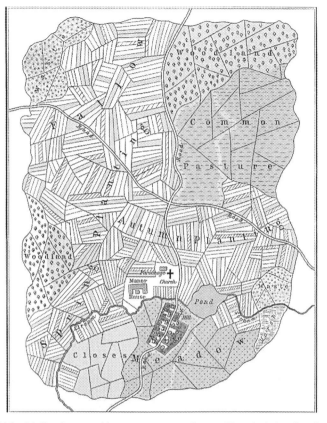

FIGURE 14.2   Medieval manor with common pastures. *Source: Map of a fictional medieval manor retrieved from <http://en.wikipedia.org/wiki/File:Plan_mediaeval_manor.jpg>.*

the behaviors of today's people who share these contemporary commons with the behaviors of fifteenth century farmers who shared common pastures. To appreciate his analogy, picture the expansive common pastures of a medieval manor (see Fig. 14.2). Although these large *commons* can support many cattle, guess what will happen if every farmer uses this land only to his own advantage.

Hardin imagined that each individual might want to graze just one more cow to improve his lot in life. By itself, this small addition does no damage, but if everyone does the same, the land is eventually ruined from overgrazing and everyone loses. The *tragedy* of these acts is not the unhappiness that comes from damage, says Hardin, but that the damage is a result of how the system works.

In systems thinking terms, Hardin might have said that this system of land-sharing behaves according to the interaction among its individual parts. Because we are locked into a system of "individual freedoms," we behave as "independent, rational, free enterprisers" (Hardin, 1968) and excuse the consequences of our actions. Counter to Adam Smith's Invisible Hand philosophy which suggests that individual acts self-regulate to benefit the common good (see Chapter 4), Hardin recognized that self-interested acts of individuals can ruin a common resource and devastate the common good. His message applies perfectly to today's sustainability issues.

14.1.1 Tragedy of the Commons and Sustainability

The notion of "tragedy of the commons" is central to sustainability and an integral part of systems thinking primarily because it describes behaviors that pit individual short-term gains against long-term losses for the group. It applies whenever individuals overuse or abuse limited shared resources on the basis of only their own needs. Using the language of systems thinking, Fig. 14.3 illustrates how limits on shared resources and others' actions relative to those resources *eventually* inhibit gains for all individuals in the group. As total use of common resources increases, in the long run, each person gains less. This behavior is so widespread that it is called a system "archetype." Sustainability-related examples range from overuse of fossil fuels and groundwater to multiple forms of pollution. An important part of this system construct is the *delay* at its center. Because personal gain is not affected immediately, individuals disregard the potential effects of their actions on the whole.

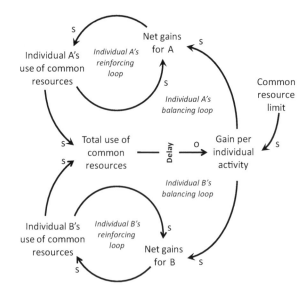

FIGURE 14.3 System archetype for tragedy of the commons. *Source: Author's depiction of CLD for tragedy of the commons; this archetype can be found in various sources including Senge (1990) and Braun (2002).*

---

3. World Commission on Environment and Development (1991) lists three; Cleveland (1993) lists four.

Experts believe the world shares four commons: (1) oceans, (2) atmosphere, (3) outer space, and (4) Antarctica.[3] Recent views declare cyberspace – the invisible domain of electronic communication – as a new commons (Bajaj, 2012). Although these few are vital, others are as important when we are talking about sustainability. Because humans are part of a world-sized integrated system, this list must include *all* land, fisheries, forests, food, water, air, and energy, regardless of where we live. Local actions have global effects. Food grown on the land and with the water from one nation is exported to other nations; pollution fouls the air, land, and water in one nation, but eventually travels around the world – even if in diluted form. Although we may think that our home country "owns" the water, oil, food, and air it uses, in the end these resources touch us all.

14.1.2   Repercussions of Mismanaging the Global Commons

We inhabit an extraordinary planet whose abundant resources support our lives; but with the help of technology and growing populations, we also have unparalleled power to destroy our inheritance. Chapter 7 proposed four resource categories that contribute to Earth's carrying capacity and its ability to sustain us: (1) energy supply (fossil fuels, uranium, and renewables); (2) water supply (ground and surface); (3) food supply (food-producing land and fisheries); and (4) forests. For this discussion, we add air quality to the list of shared resources and consider these five as our global commons. Without these bare necessities, life would be compromised.

The systems challenge to managing this global commons is that the effects of individual behaviors do not show themselves for decades or even centuries. Many actions do not seem to matter at the individual level – use of limited water and fossil fuels, economic growth, and excessive waste – but over time they lead to tragedy for our global commons. The integrated actions of collectives and individuals discussed in Chapter 13 now take on new meaning. To preserve these shared resources, we must transcend personal ownership before the accumulation of individual self-motivated behaviors has literally eaten us out of house and home.

Now, with this final perspective about the global commons planted firmly in our minds, we can compose a list of primary insights gained by applying systems thinking to sustainability.

14.2   INSIGHTS FROM SYSTEMS THINKING

This book began with the premise that sustainability is a complex topic that is impossible to understand without viewing its interdependencies. It recommended we use systems thinking to understand the effects of our actions and the potential outcomes from narrowly devised plans for change. Now that we are at the end of the book, we can say that by using systems thinking, we did get a better appreciation for the relationships and dynamics that define sustainability. In addition, systems thinking has given us three encompassing insights that summarize the whole of what we have learned.

14.2.1 Inertia, Balance, and Perspective

We now know that many of the beliefs we hold about how the world works – our mental model – is incomplete and is leading us down a primrose path. An integrated system diagram emerged from our analysis to help us visualize sustainability's intricate web of relationships. For convenience, Fig. 14.4 repeats the system diagram from Chapter 8.

In addition to appreciating this remarkable network of cause and effect, the first important lesson from applying systems thinking to sustainability is that the system's *inertia* (its resistance to change) is immense. It is like a runaway freight train on a downhill slope that we are trying to stop with a two-inch wall of white styrofoam. Inertia in the sustainability system comes from its numerous reinforcing loops that generate ever-accelerating growth and from its weakened balancing loops that cannot slow the growth machine down. Multiple lengthy delays in these balancing loops prevent system limits from taking immediate effect. These relationships also dictate that inertia will be present if we have triggered collapse and the system is decaying. In other words, once growth or decay starts, it takes time to stop. Our actions must first slow the train down before we can stop it and before it crashes.

The second lesson, which we have reiterated throughout the book, bears repeating. This lesson reminds us of the need for *balance*. Balance requires that we give equal weight to conflicting considerations and helps overcome inertia as we transition toward a sustainable world. It is the tightrope we must walk until we have decreased excessive growth and the system reaches a state in which competing forces work together in harmony – until we achieve sustainability.

We can tell from the system diagram that to attain balance, tradeoffs among proposed solutions are crucial. Because we are butting up against the Earth's ability to support the billions of humans who are alive today and the billions more who will be alive in the next decades, our population, economy, instant gratification, energy consumption, and pollution cannot continue to grow; we cannot have it all. Emphasizing one area will push the others off-kilter and cause our survival to wobble like a red top on its last few spins. Sustainability requires many types of balance: balance among economy, society, and environment; balance between present and future; balance in the ways we achieve happiness and well-being; and balance between our small selves and the welfare of society.

The importance of *perspective* is the ultimate lesson. The systems perspective featured in this book opened our eyes to view sustainability in its broadest context and to understand how its major components influence one another. The resulting system diagram is a framework we can use to introduce new factors. I encourage you to think of this diagram when you read or hear about sustainability issues. Superimpose them on the diagram's elements and arrows of influence to see what happens. Even though we have not nearly covered all the weighty topics that rest on sustainability's broad shoulders, a systems perspective helps us think about how our decisions, actions, and behaviors affect the future.

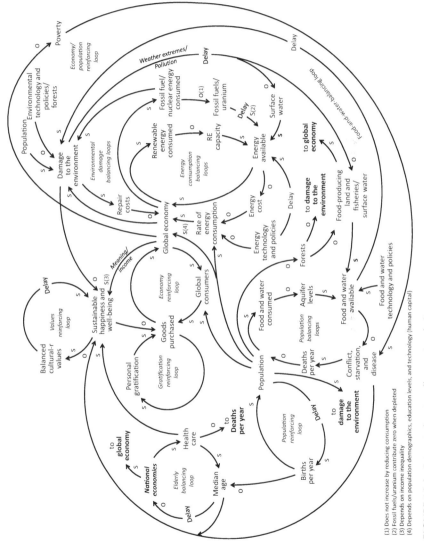

FIGURE 14.4 Integrated system diagram (reprise).

(1) Does not increase by reducing consumption
(2) Fossil fuels/uranium contribute zero when depleted
(3) Depends on income inequality
(4) Depends on population demographics, education levels, and technology (human capital)

14.2.2 Systems Thinking and Effective Intervention

Chapter 9 proposed a plan of intervention for sustainability. The plan takes the three lessons of systems thinking to heart by dividing the areas of intervention into three levels (paradigm shifts, structural changes and transition to the future. Within these levels, it first acknowledges the massive *inertia* in the system and includes actions that gradually build a conceptual foundation, actions that reduce population and pollution to relieve currently rising pressures, and actions such as technology investment that recognize the power of inertia and give us time to adjust today's unsustainable behaviors.

Second, the plan proposes a *balance* between present and future, between individual happiness and collective well-being, between national sovereignty and world concerns, and among the three components of sustainability. In achieving this balance, it emphasizes harmony, integration, and simultaneous action, accommodates human nature, and proposes ways to engage autonomous individual commitment within the structural confines of global governance.

Finally, the plan incorporates *perspective*. It considers issues and solutions as they relate to one another and makes tradeoffs among them. It causes us to fast-forward in time so that we can understand and accommodate the effects of present actions. With its systems framework, the plan releases us from the confines of our mental model. It addresses major aspects of sustainability and proposes actions that touch every person on Earth. Finally, it adds the notion of sustainable stewardship to integrate individual and collective efforts.

14.3 WHERE TO NOW?

For a healthy tomorrow, today's society must tackle the disconcerting challenge of keeping Mother Earth alive and well. Our health is tied to her health. Even with the holistic insights from systems thinking, success depends on each and every one of us. It depends on our cultures and on our social institutions. Do individuals, organizations, and world leaders have the courage to change? Can we overcome reliance on growth and addiction to consumption regardless of whether it is energy and water or beef and bling? Can we change our lifestyles and our beliefs? Can we live within Earth's carrying capacity?

We must certainly give it a try. We can no longer go about our short-term business, measuring success in terms of economic growth, personal wealth, and buying things. We must finally burst our mental bubble and adapt to a new reality. We have no choice, for as system dynamics expert Jay Forrester says "if we follow intuition and the fallacies embedded in mental models, the trends of the past continue into deepening difficulty" (Forrester, 1971). Our new reality moves us toward sustainable stewardship.

Indeed, we humans live in a most uncommon world. We do not have all the answers, but we do know that managing the abundant resources in our great global commons is the heart of sustainability. There are inklings that we are on the right path as we try to turn Garrett Hardin's "tragedy of the commons" into what social theorist Jeremy Rifkin calls the "rise of a collaborative commons" in which the sharing of goods and services reshape how we think about economic performance (Hardin, 1968; Rifkin, 2014). In this new commons, we translate our concerns for community and future into practical action; we unite our efforts rather than sit back and expect a common good to appear from acts of self-interest.

Perhaps one day we will realize that our paradigms about the world must continually be reexamined for every paradigm represents our "limited understanding of an immense and amazing universe that is far beyond human comprehension" (Meadows, 2008). Perhaps one day humanity will arrive at the spiritual understanding of life that Buddhists call "Enlightenment." Perhaps one day we will have global governance for sustainability which, like Star Trek's Federation, will have its own prime directive: "Live within Earth's ability to sustain life!"

In the meantime, and until we find better answers, we must take our global citizenship and our stewardship role seriously. By depending on technology to cure the Earth, or letting sporadic initiatives blind us to our responsibilities, or burying our heads while we continue our patterns of excess, we are avoiding the issues and moving rapidly away from sustainability.

We can learn what *not to do* from the ancient Mayan civilization that failed to adapt to its environment or from the great Roman Empire that outgrew its resources. We can learn what *to do* by combining our thoughts and efforts, our failures, and successes in global debates and discussions. We can view sustainability as a big system whose current behavior will be ponderously slow to change and whose interacting parts have both short-term effects and long-term consequences. In these ways, we will recognize the urgency for change and equip ourselves to identify solutions and consequences.

My hope is that you have deepened your understanding by reading this book and are inspired to begin your personal journey. My wish is that you will encourage others to begin their own journeys.

If Earth could speak, its sad voice would haunt us with pleas for our help. We must listen. We must acknowledge that while we cannot do it alone, we each have a part. Let us appreciate the preciousness of this jewel that we call home – our unique, irreplaceable, and uncommon globe.

REFERENCES

Bajaj, K., April 16, 2012. Cyberspace as global commons: the challenges. DATAQUEST. Retrieved from <http://www.dqindia.com/dataquest/news/142213/cyberspace-global-commons-the-challenges>.

Braun, W., 2002. The System Archetypes. Retrieved from <http://www.albany.edu/faculty/gpr/PAD724/724WebArticles/sys_archetypes.pdf>.

Cleveland, H., 1993. Birth of a New World. Jossey-Bass, San Francisco.

Forrester, J., 1971. Counterintuitive behavior of social systems. Technology Review. Cambridge: Alumni Association of the Massachusetts Institute of Technology Updated March 1995. Retrieved from <http://clexchange.org/ftp/documents/system-dynamics/SD1993-01CounterintuitiveBe.pdf>.

Hardin, G., 1968. The tragedy of the commons. Science 162, 1243–1248.

Meadows, D.H., 2008. Thinking in Systems: A Primer. Chelsea Green Publishing, White River Junction, VT.

Rifkin, J., 2014. The Zero Marginal Cost Society: The Internet of Things, the Collaborative Commons, and the Eclipse of Capitalism. Palgrave Macmillan, New York.

Senge, P., 1990. The Fifth Discipline: The Art & Practice of the Learning Organization. Doubleday/Currency, New York.

White, F., 1987. The Overview Effect: Space Exploration and Human Evolution. Houghton-Mifflin, Boston.

World Commission on Environment and Development, 1991. Our Common Future. Oxford University Press, Oxford.

Glossary

Aquifer Aquifers are underground stores of water-bearing permeable rock, gravel, silt, and sand. They are found at different levels, the shallowest of which are easily replenished by precipitation. Significant aquifers that were formed millions of years ago (called "fossil water") lie deep under the Sahara Desert in northern Africa, in the Guarani Aquifer in South America, under the Kalahari Desert in southern Africa, in the Great Artesian Basin in Australia, in the High Plains Aquifer under the Midwestern United States (which includes the Ogallala Aquifer) and various other places, including beneath the oceans. In this book, we consider aquifers that are replenished little, if any, by surface water to be limiting factors on Earth's ability to sustain us.

Balance Rather than a physiological characteristic that keeps individuals upright, balance as used in this book complements the philosophy of systems thinking. Here, balance refers to a situation in which elements have equal importance and work together in harmony. It can apply to conditions in which there are two equally beneficial, but conflicting considerations. Such considerations include present versus future; individual versus collective; nation versus world; and the competing demands of economy, society, and environment. For healthy living systems such as the human body or Earth's ecosystem, the desired state is one of dynamic balance in which system components adapt harmoniously to change. For sustainability, we describe it as the condition in which Earth's carrying capacity can continuously provide a reasonable quality of life for its current and future inhabitants.

Carrying Capacity Carrying capacity refers to the ability of an environment to sustain the lives of the organisms that rely on it for support. Systems thinking uses the concept of carrying capacity to describe a system's limits and its ability to sustain growth. Sustainability of Earth depends on its carrying capacity, that is, its ability to keep its population alive and well. In this sense, carrying capacity incorporates various limiting factors such as energy, food and water supplies, clean air, and forests.

Causal Loop Diagram In systems thinking terms, causal loop diagrams are simplified ways to describe essential elements and relationships in a system. These diagrams include curved causal-link arrows (depicting influence from cause to effect) and the polarity of that linkage. An "s" indicates that cause and effect move in the *same* direction and an "o" shows that they move in *opposite* directions (e.g., when cause increases, effect decreases below what it would have been). Causal-link arrows combine into balancing (B) and reinforcing (R) feedback loops. Significant lags between an action and the effects of that action appear as "delay" on the causal-link arrows.

Elderly Dependency Ratio The elderly dependency ratio (EDR) calculates the proportion of "elderly" people to "working age" individuals. Its specific formula is: EDR = (people aged 65 and over/people aged 15–64) × 100. It compares the dependent part of society with the productive part as a function of age. EDR rises when, the number of elderly grows faster than the number of working age people

K.L. Higgins: Economic Growth and Sustainability. http://dx.doi.org/10.1016/B978-0-12-802204-7.00016-5

Feedback Loops Feedback loops are a systems thinking construct that represent a continuous process in which the effect of an action comes back to influence the original cause of that action. There are two basic types of feedback loops: (1) balancing loops and (2) reinforcing loops. Balancing loops tend to stabilize a system and bring it toward a goal or limit. Reinforcing loops are engines of growth or decay and push the system the way it is already going.

Fertility Rate Different from birth rate, which measures the number of births in a particular region over a particular period of time, fertility rate indicates the average number of children that *would* be born to a woman *if* she were to experience the age-specific fertility rates that predominate in her country. Fertility rates are an indicator of the social, cultural, and economic factors in a nation or region.

Fixes That Fail In systems thinking terms, fixes that fail is a common behavior called an "archetype". It describes a situation in which a solution that is effective in the short term has unintended consequences that appear after some delay to exacerbate the original problem. The causal loop diagram for this archetype begins with a balancing loop to fix the short-term problem and adds a reinforcing feedback loop to incorporate the delayed effect of that fix.

Gini Index for Income Inequality The Gini index or Gini ratio, named after Italian statistician Corrado Gini, measures the inequality of given variables. For income, it describes the concentration of household income in a country. This index ranges from 0.0, where every household has the same income, to 1.0 where one household has all the income.

Global Commons Traditionally, the global commons refers to Earth's common and shared resources such as oceans, atmosphere, outer space, and Antarctica. More recently, some have added cyberspace (our shared communications environment) to this list. In this book the global commons for sustainability includes energy, food and water supplies, forests, and air quality. Management of our global commons is a frequently visited topic that continues to challenge us.

Governance Unlike governments that influence and direct groups of people through laws and statutes, the concept of governance refers to a combination of formal laws and policies, and informal influences through the culture, norms, and values of social units such as families, social networks, and organizations. Governance is a less directive approach to shaping behaviors and requires unity of values and goals.

Gross Domestic Product (GDP) First developed in 1934, this economic metric represents the output of goods and services produced by all labor and property within a country for some stated period.

Leverage Leverage is a powerful concept for systems thinking. It is the key to finding long-lasting solutions that directly affect major elements in the system. A leveraging action is one in which small efforts yield big results; in other words, leverage amplifies efforts. The most effective leveraging actions in a system involve paradigm shifts and structural changes that alter relationships, the strength of feedback loops, or delay times. Identifying leveraging actions is a significant benefit of applying systems thinking to a complex issue.

Limits to Growth In systems thinking terms, limits affect growth and decay. When a system depends upon a certain resource, it can grow as long as that resource is available, that is, within the bounds of its capacity. When the system has grown so much that it has depleted its resources, it has hit its limits and growth will halt. Balancing feedback loops create the primary mechanism that slows down or stops growth from reinforcing loops. In systems thinking terms, limits to growth is depicted by a reinforcing loop coupled with one or more balancing loops whose strength increases as the system approaches its limits. This system archetype appropriately applies to sustainability, particularly as Earth's resources such as energy and water are depleted. When a system reaches or exceeds its limits, it can either smoothly stop growing or hit a peak and collapse.

Median Age Median age is the age at which half a given population is older and half is younger.

MTOE MTOE stands for a million tonnes of oil equivalent – a metric for fossil fuel and other types of energy consumption.

OECD The Organisation for Economic Co-operation and Development was established in 1961 to improve the well-being of the world's people. Its 34 member countries work together to identify and analyze world problems and to promote economic and social policies to solve them.[1] Member countries include advanced economies as well as Mexico, Chile, and Turkey; OECD works closely with emerging nations like China, Brazil, and India.

Purchasing Power Parity Purchasing power parity (PPP) is a normalizing factor used to express gross domestic product of a nation. It reflects the cost of similar items in different countries so that their GDPs can be compared.

Reinforcing Feedback Loops Reinforcing feedback loops are systems thinking constructs that reflect continuous, often exponential, growth or decay.

Sustainability Sustainability is the biological principle that allows organisms to remain diverse and productive, and to endure over time. Its organizing tenet is sustainable development, which the World Commission on Environment and Development formally defined as "development that meets the needs of the present without affecting the ability of future generations to meet their own needs" (World Commission on Environment and Development, 1991). For clarity, this book applies the latter definition to the concept of sustainability. As its framework, it uses three overlapping domains of economy, environment, and society. These domains were defined at The World Conservation Congress in Bangkok, Thailand, November 17–25, 2004 (IUCN, 2004). Some versions of sustainability's components also include culture and education. One inherent dilemma of sustainability is that it pits present against future.

System Dynamics System dynamics was derived from systems theory whose concepts originated in the 1700s. System dynamics became popular in the 1950s and 1960s. It uses models and computer simulations to understand behavior of an entire system, and has been applied to the behavior of large and complex national issues. It portrays the relationships in systems as feedback loops, lags, and other descriptors to explain dynamics, that is, how a system behaves over time. Its quantitative methodology relies on what are called "stock-and-flow diagrams" that reflect how levels of specific elements accumulate over time and the rate at which they change. Qualitative systems thinking constructs evolved from this quantitative discipline.

Systems Thinking Systems thinking is a discipline or process that considers how individual elements interact with one another as part of a whole entity. As an approach to solving problems, systems thinking uses relationships among individual elements and the dynamics of these relationships to explain the behavior of systems such as an ecosystem, social system, or organization.

Tragedy of the Commons First described by Garrett Hardin in 1968 (Hardin, 1968), tragedy of the commons is a common social dilemma, particularly for systems that involve

---

[1]See <http://www.oecd.org/about/history/>.

human beings. It reflects system behavior when individuals who share common, limited resources act in their own best interest: Self-interested acts eventually harm the common good. Each individual believes that his or her small acts do not matter; in the aggregate they can create disastrous situations. In systems thinking terms, tragedy of the commons is depicted by reinforcing loops for individuals whose gain increases with their self-interested acts. Balancing loops aggregate these individual acts and eventually diminish the benefit for individuals and deplete the common resource. For sustainability, tragedy of the commons applies to Earth's population who share limited resources such as fossil fuels and water; it is also one reason that pollution grows so rapidly.

REFERENCES

Hardin, G., 1968. The tragedy of the commons. Science 162, 1243–1248.

IUCN, 2004. The IUCN Programme 2005–2008: Many Voices, One Earth. Adopted at The World Conservation Congress, Bangkok, Thailand, November 17–25, 2004. Retrieved from <http://cmsdata.iucn.org/downloads/programme_english.pdf>.

World Commission on Environment and Development, 1991. Our Common Future. Oxford University Press, Oxford.

Index

A

Agricultural waste, 87
Air pollution, 61, 98
Aquifers, 83, 104

B

Balance, need for, 186, 188
Balanced cultural values, 118, 126, 129
　new economics and new metrics for, 132
　society's metrics for success, 132
　for sustainable happiness and well-being,
　　121
Balancing feedback loops, 6, 90, 96
　energy consumption, 98
　fossil fuel/nuclear energy, 98
　population, 102
Beer, Stafford, 9
Beliefs of mental models, 16
Beneficial balance between economy and
　　society, values and, 105
Birth rate. *See also* Population growth
　family planning and, 54
　interventions for controlling, 139
　　China's one-child policy, 141
　　education and indoctrination for,
　　　139
　　education levels, 140
　　national policies on population
　　　control, 140
　　policies for population control, 140
Brundtland Commission, 1983, 3
Brusaw, Julie, 176
Brusaw, Scott, 176

C

Cameron, James, 95
Carbon-based energy resources, 16, 156
Carbon dioxide (CO_2) emissions, 57, 58
　countries with high, 151
　deforestation and, 88
　efforts to manage, 152, 154
　global surface air temperature anomalies
　　and, (1965–2040), 58

Carbon price, 22
Carley, Michael, 16
Carrying capacity of a system, 76, 96
　determinants, 76
　Earth's, 76, 77, 90, 98, 108, 139, 147,
　　185, 188
　mental model and, 77
Carson, Rachel, 1
Causal loop diagrams, 4–6
Christie, Ian, 16
Clean Air Act (1970), 56, 154
Climate engineering, 154
Club of Rome, 1968, 3
Coal consumption, 79
Collaborative commons, 188
Consumer spending and economic growth,
　　33, 34
Cultural values
　balanced, 118, 126, 129
　economic growth and, 33
Cushman, Philip, 17

D

Death rate. *See also* Population growth
　1950–2013, 54
Deforestation, 88, 105, 150, 151, 154
Delays, 6, 8, 9, 69, 77, 100, 106, 110, 116,
　　118, 184, 186
Diamond, Jared, 29
Dichter, Ernest, 46
Drought, impact of, 84

E

Earth Day, 3
Earth jurisprudence, 172
Earth's ecosystem, carrying capacity of, 90
Eastern philosophies, 42
　values, 105
Economic growth, 28
　Adam Smith's Invisible Hand philosophy,
　　168, 183
　addictions and, 32, 38
　aging and, 55

Economic growth *(cont.)*
   collectives and economic success, 17
   cultural reasons for, 33
   environmental damage and, 97–99
   fuel consumption and, 79
   gross domestic product (GDP) and, 17, 18,
      33, 34
   happiness and, 38
   health care, benefits of, 103
   implications for standards of living, 35
   individuals and materialism, 16
   limitations of, 168
   personal gratification and, 26
   personal reasons for, 32
   population and consumer spending,
      33, 34
   reinforcing loops and, 26
   repercussions of, 37
   short-term benefits, 36
   societal reason for, 33
   technology advances and, 18, 27, 31
   trends, 33
   world trend, 18
Economy/population reinforcing loop, 120,
    136, 149
Economy-reinforcing loop, 26, 27, 97, 98, 120,
    127, 136, 149, 187
Egyptian Old Kingdom, 10
Egyptian ruins, 2
Elderly balancing loop, 103, 119, 120, 127,
    136, 149, 187
Elderly dependency ratio (EDR), 54,
    55, 143
Elderly people, 55
   public spending for, 56
Emerson, Ralph Waldo, 1
Energy, 18, 27, 29
   availability of, 19
   consumption, 19, 78, 79, 121
   cost, 137
      preparing for cost increase, 137
      raising, 138
      recovery of full, products and services,
        137
   fossil fuel, 19, 78, 98
   green, 19, 157
   hydroelectric, 19, 78, 81, 82, 99
   nuclear, 80, 81, 98
   other renewable energy, 78, 81, 82
   -powered devices, 19
   -related technology, 19
   renewable, 81
   solar, 20

Energy consumption-balancing loops, 97, 98,
    120, 136, 149, 187
Energy consumption-reinforcing loop, 27, 98
Energy cost, 99
Energy returned for energy invested (EROEI),
    79
Energy technology and policies, benefits of, 99
Environmental damage and economic growth,
    97–99
Environmental damage balancing loops, 97,
    119, 120, 127, 136, 149, 187
Environmentalism, 2
Environmental Protection Agency, 3
Environmental technology and policies,
    benefits of, 99

F
Feedback loops, 6, 135
   altering, 135
Fertility rates, 51, 53
   between 2015 and 2040, 51
   countries with low, 141, 144
   in Iran, 141
   of selected countries (1991 and 2011), 53
Fisheries, 85
Fixes that fail, 148
Food and water balancing loop, 104, 119, 120,
    127, 136, 149, 187
Food-producing land and fisheries, 85, 105
   better farming methods, 160
   increasing, 160
   technology solutions and policies, 158
Food supply, limitations of, 85
   changes in food production, 86
   food-producing land and fisheries, 85, 105
   future limits, 86
Forest preservation, 89, 154
Forrester, Jay, 9, 14, 188
Fossil fuel emissions, efforts to manage, 152
Fossil fuel energy, 19
   annual consumption of, 78
   balancing loop for, 98
   limiting factor of, 78
Fracking (hydraulic fracturing), 20, 21
Frankl, Victor, 45
Franklin, Benjamin, 138
Fukushima Daini nuclear plant tragedy, 64
Fullan, Michael, 128

G
Gaia Hypothesis, 95, 96, 110
Gini index for income inequality, 43, 44

Gleick, Peter, 161
Global "buy-in" of a sustainability plan, 172
Global collaboration, 148, 171
Global commons, 181, 182
    repercussions of mismanaging the, 185
Global consumers, 98
Global consumption of goods and services, 35
Global economy, 26, 27, 98, 102, 148
Global governance, 171
    for sustainability, 172
Global warming, 57, 64, 84
Gowdy, John, 168
Grass-roots organizations, role in sustainable
        stewardship, 175
Gratification-reinforcing loop, 26, 27, 41, 97,
            98, 105, 120, 127, 136, 149, 187
Gravelet, Jean-François, 113
Greatest Happiness Principle, 16
Great London Smog (1952), 56
Green energy and energy conservation, 19, 157
Greenhouse gases (GHG), 57, 98, 150
    Arctic and Antarctic ice loss, 58
    carbon dioxide (CO_2) emissions, 57, 58
    contribution to global warming, 57, 58
    efforts to manage, 151, 152, 154
    fluorinated gas emissions, 57
    methane emissions, 57
    nitrous oxide emissions, 57
    by sector, 151
Greenpeace, 171
Gross domestic product (GDP), 17, 18, 33, 34,
            36, 67, 138, 174
    in Brazil, 35
    in China, 35
    in Mexico, 35
    in Nigeria, 36
    in South Sudan, 35
Gross national happiness, 18
Groundwater depletion, 83
    water conservation efforts for, 83
        developing new water sources, 84
G8 world forum, 171
G20 world forum, 171

H

Happiness/well-being, 42, 44, 100, 104, 148
    balanced cultural values and, 121
    carbon dioxide (CO_2) emissions, impact
        on, 60
    health care and, 46, 47
    increased longevity and, 45
    link between money and, 42

physical fitness and mental health, effect
    of, 45
positive emotions, 45
relationships, health, and meaning, 44
short-term/long-term trade-offs of, 46
sustainable, 42, 100, 104, 148
    balanced cultural values and, 121
    system diagram, 105
    system diagram, 105
Happy Planet Index, 33
Hardin, Garrett, 29, 168, 182
Health care, benefits of, 103
Heinberg, Richard, 42
Hempel, Monty, 177
Hobbes, Thomas, 15
Hydroelectricity, 81

I

Income inequality, 42
Individual contributions for sustainable
        stewardship, 176
    individual actions, 178
    self-initiated individual contributions, 176
    understanding human nature and behavior
        change, 177
Individual mental models, constraints of, 15
Industrial, agricultural, and human wastes, 65
    Ganges River Basin, case of, 65
Industrial Age, 1760s, 56
Inertia, 186
Integrated system, 187
    diagram, 101
        mental model and, 106
    sustainability as an, 9, 95
    transition of mental models to, 169
    transition to sustainable stewardship, 170
        collective component of, 171
        formal courses and educational
            programs, 174
        global collaboration and global
            governance, 171
        grass-roots organizations, role of, 175
        individual contributions, 176
        media, role of, 174
        millennial generation, impact of, 175
        potential sources of influence, 173
        transnational corporations (TNCs), role
            of, 174
Integration, 95, 96, 110, 162, 170, 188
Interdependence, 1, 4, 69, 95, 96, 110, 162, 169
International Union for Conservation of
        Nature, 3

Interventions for sustainability. *See also*
Technology solutions and policies
altering feedback loops, 135
areas of, 118, 119, 127, 130
energy cost, 137, 187
generic types, 115, 116
involving environmental repair, 153
reducing GHG, 154
median age and aging population, 142
retirement related changes, 143, 144
reduction of births per year, 139
education and indoctrination for, 139
national policies on population control,
140
three-level framework for, 170

K

Klein, Stefan, 45
Kyoto Protocol, 22, 138

L

Lags. *See* Delays
LeDoux, Joseph, 15
Leverage for sustainability, 114, 115
Life expectancy, 45, 103
Limiting factors, 77
diminishing Earth's ability to clean
atmosphere, 88
energy supply, 78
fossil fuels, 78
nuclear energy, 80
renewable energy, 81
food supply, 85
changes in food production, 86
food-producing land and fisheries, 85
future limits, 86
forests, 88
deforestation impact, 88
future limits, 89
tropical rainforests, 88
system diagram, 100
water supply, 82
future limits, 84
groundwater, 82, 83
surface water, 83
Limits-to-growth, 7, 29
ancient civilization and, 73
Lovelock, James, 95

M

Malthus, Thomas, 50
Man's Search for Meaning (Victor Frankl), 45

Margulis, Lynn, 95
Material consumption, 16
Mayan civilization, decline of, 10
Mayan ruins, 2
Meadows, Donella, 9, 115, 116
Media, role in sustainable stewardship, 174
Median ages, 54, 102, 103, 106
in 2010, 55
interventions for, 142
balancing among nations, 143
retirement related changes, 143, 144
Mental models
arousing moral commitment, 128
constraints of individual mental models,
15
definition, 14
development of, 14, 169
dissemination of information in multiple
formats and media, 128, 129
effects of actions, 126
energy and technology advances,
significance of, 18
implications, 28
plastic bubble analogy, 13, 168
population growth and, 21
primary beliefs, 16
systems depiction of, 25
vs unrealistic beliefs, 29
Milankovitch cycle, 57
Millennial generation, role in sustainable
stewardship, 175
Million tonnes of oil equivalent (MTOE), 58,
78, 107
Mortality rate. *See* Death rate
Muir, John, 1
Municipal solid waste (MSW), 61
E-waste dumpsite in Ghana, 63
generation per capita by region (2012 and
2025), 62

N

Nested networks, 157
"New economics" movement, 132
Nuclear capacity, 80
Nuclear energy, limits on, 80
balancing loop for, 98
nuclear power generation, 81

O

Obesity statistics, 159
Ogallala aquifer, 83, 84
Oil consumption, 79

Oil production, between 1990 and 2013, 80
Old Kingdom of ancient Egypt, 1
Organization for Economic Co-operation and
        Development (OECD), 159
    aggregate household spending, 18
    obesity data from, 159
    waste per capita, 61, 62
Overshoot-and-collapse, 8–9, 130, 168

P

Personal gratification, 41, 98
Perspective, importance of, 186, 188
Pollution, 22, 28, 38, 56, 85, 98, 168–169
    air, 61
    greenhouse gas (GHG), 57
    industrial, agricultural, and human wastes, 65
    intervention for, 150 , 153–154
    municipal solid waste (MSW), 61
    primary pollutants, 56
    radioactive waste, 63
    repercussions of, 56
Population balancing loop, 119, 120, 136, 149,
        187
Population-balancing loop, 102
Population decrease, consequences, 108
Population growth, 21, 38, 50, 85, 107, 121.
        See also Birth rate; Death rate
    1940–2040, 52
    2015–2030, 107
    aggregated statistics, 51
    between 1965 and 2014, 53
    between 2015 and 2040, 51
    birth and death rates, 53, 54
    challenges, 53
    consequences of, 66
    elderly dependency ratio (EDR), 54, 55
    environmental damage and, 104
    fertility rates, 51, 53
    impact on economy, 102
    intervention for, 139
    median ages, 54
    policies for population control, 140
    post Industrial Age, 50
    proportion of "elderly" people, 55
    replacement rate, 51
    response to crowded conditions, 66
    selected food production and arable land
            vs, 1965–2010, 86, 87
    spread and concentration of pollution and,
            22, 28
    trends, 1965–2040, 68
    "working age" individuals, 55

Population reinforcing loop, 120, 136, 149, 187
Population-reinforcing loop, 102, 106
Positive emotions, 45
Positive experience index, 45
Powers, Bill, 79
Purchasing Power Parity (PPP), 58
    GDP in terms of, 36
Puzzles, 125

R

Radioactive waste, 63, 98
Reforestation, 154
Reinforcing feedback loops, 6, 90
    economic growth and, 26
    economy, 26, 98
    energy consumption, 27
    gratification, 26, 27, 41, 98, 105
    population, 102, 106
    system diagram, 102
    values, 106, 131
Renewable energy, 81, 157
    Policy Network, 157
Rifkin, Jeremy, 188
Rio +20 agenda, 173
Ryerson, William, 53

S

Salinization, 87
Schmookler, Andrew, 31, 126
School education, sustainability practices in, 174
Schweikart, Russell, 181
Senge, Peter, 5
Simultaneous Policy (Simpol) campaign, 175
Skinner, B.F., 15, 19
Smith, Adam
    Invisible Hand philosophy, 168, 183
Solar power, 20, 158
Solar Roadways project, 176
Spilsbury, John, 125
S-shaped growth, 7, 8
Standards of living and economic growth, 35
Staudt, Amanda, 60
Sterman, John, 9, 76
Stiglitz, Joseph, 44
Super-cycle, 35
Sustainability, 89, 90, 110, 168
    analytic approach to, 115
        assigning dates, 118
        creating synergy, 118
        defining actions, 118
        finding candidate areas, 117

Sustainability *(cont.)*
    finding responsible parties, 118
    prioritizing areas of intervention, 117
    validating feasibility, 117
  as an integrated system, 9, 95
  defined, 2
  dynamic interdependencies of, 4, 95
  foundation building for, 126
  global governance for, 172
  importance of perspective, 185
  inertia in, 185
  influencing factors, 148
  intervention areas and actions to, 120, 122
  long-term, 113
  need for balance, 185
  overlapping circles of, 3, 169
  proposed actions, 121
    levels of effectiveness, 121
    opposing effects, 121
  related beliefs, 25, 26
  solutions, 114
    areas of intervention, 118, 119, 127
    generic types of intervention, 115, 116
    leverage, 114, 115
  synergistic plan to achieve, 125
Sustainability revolution, 2
Sustainable agriculture and water conservation,
    161
Sustainable civilization, 10
Sustainable collaboration, 173
Sustainable development, 2
Sustainable happiness and well-being, 42, 100,
    104, 148
  balanced cultural values and, 121
  system diagram, 105
Sustainable stewardship, 188
  grass-roots organizations and, 175
  individual contributions for, 176
  media and, 174
  millennial generation, role in, 175
  transition of integrated system, 170
    collective component of, 171
  transnational corporations (TNCs) and, 174
System boundaries, 7
System diagram, 96
  areas of system imbalance, 109
    emphasis among economy,
      environment, and society, 110
    self-interest vs community interest,
      110
    short-term vs long-term, 109
  balancing loops, 96, 98
  carrying capacity, 96

  cultural values, sustainable happiness, and
    well-being, 105
  damage to environment, 104
  delays, 100, 106
  economic growth and environmental
    damages, 98
  economy/population-reinforcing loop, 102
  elderly balancing loop, 103, 119, 120, 127,
    136, 149, 187
  energy consumption, 98
  energy consumption-reinforcing loop, 27, 98
  environmental damage balancing loops, 97,
    119, 120, 127, 136, 149, 187
  focal points for change in, 135
  food and water balancing loop, 104, 119,
    120, 127, 136, 149, 187
  gratification-reinforcing loop, 26, 27, 41,
    97, 98, 105, 120, 127, 136, 149, 187
  integrated, 101
    mental model and, 106
  integrating economy and environment, 97,
    101
  limiting factors, 100
  partial, 97
  population dynamics, 102
    population-balancing loop, 104
    population-reinforcing loop, 104
  societal interdependencies, 100
  structure-related policies for energy cost,
    birth rate, and median age, 135, 136
  values-reinforcing loop, 105, 106, 131
  for visualizing consequences and discovering
    hope for the future, 127, 128
System dynamics, 4, 9, 188
Systems thinking, 4, 148
  application of, 9
  boundaries and limits, 7
  causal loop diagrams, 4–6
  constructs, 5
  delays, 6
  effective intervention and, 186
  feedback loops, 6
  fixes that fail, 148
  importance of perspective, 185, 188
  inertia in sustainability system, 185, 186
  insights from, 185
  need for balance, 185, 188

T

Technology advances, 18, 27
  Internet, 33
  personal computers, 33

Technology solutions and policies, 148
    energy, 152, 156
        for global conservation, energy usage, and pollution, 157
        green energy and energy conservation, 157
        renewable, 157
    for environment, energy, food, and water, 148, 149
    environmental, 150
        reduction pollution, 150
    food, 158
    for long-term vs short-term conflict resolution, 148
Thoreau, Henry David, 1
Toffler, Alvin, 3
Tragedy of the commons, 182, 188
    sustainability and, 184
Transnational corporations (TNCs), role in sustainable stewardship, 174
Tree of Souls, 95

V

Values-reinforcing loop, 105, 106, 131
Vernadsky, Vladimir, 1, 95

Vetter, David, 13
Virtual water, 84

W

Waste water management, 161
Water conservation, 160
    identifying new sources, 161
    sustainable agriculture and, 161
Waterlogging, 87
Water pollution, 21, 65
    in Indonesia, 61, 62
    Jakarta's polluted rivers, 61
Water supply, decreasing, 82
    future limits, 84
    groundwater, 82, 83
    surface water, 83
    water conservation efforts for, 83
        developing new water sources, 84
Western philosophies, 41
World Health Organization, 171
World Resources Institute's ACT 2015, 156

Z

Zemlya, Mat, 95

Printed in the United States
By Bookmasters